ISBN-13: 978-1-951619-08-4

Also Available from Dr. Steve Warner

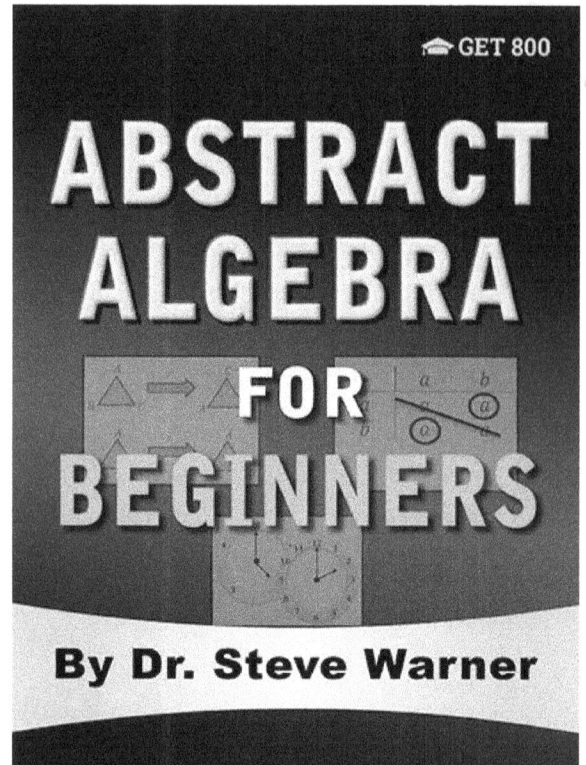

CONNECT WITH DR. STEVE WARNER

www.facebook.com/SATPrepGet800

www.youtube.com/TheSATMathPrep

www.twitter.com/SATPrepGet800

www.linkedin.com/in/DrSteveWarner

www.pinterest.com/SATPrepGet800

Also Available from Dr. Steve Warner

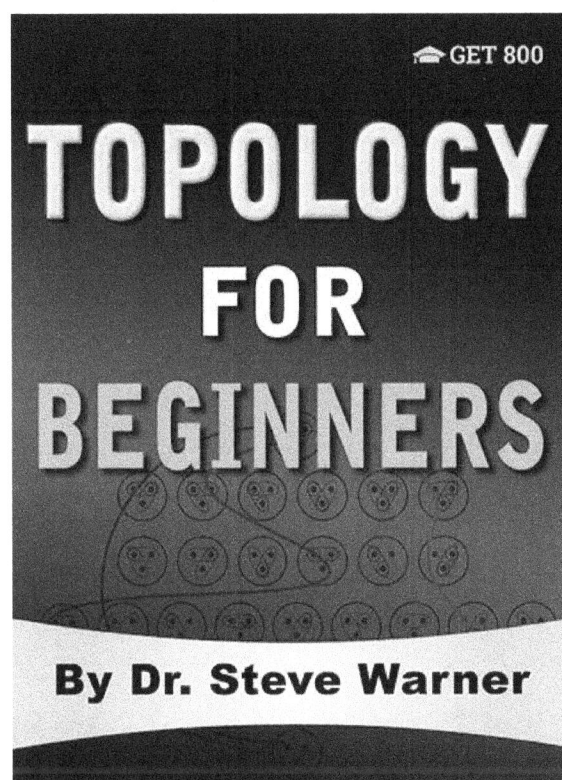

CONNECT WITH DR. STEVE WARNER

www.facebook.com/SATPrepGet800

www.youtube.com/TheSATMathPrep

www.twitter.com/SATPrepGet800

www.linkedin.com/in/DrSteveWarner

www.pinterest.com/SATPrepGet800

Prepping for Pure Mathematics

A Starter's Guide to Logic, Set Theory, Abstract Algebra, Number Theory, Real Analysis, Topology, Complex Analysis, and Linear Algebra

Dr. Steve Warner

Table of Contents

PREPPING FOR PURE MATHEMATICS

Shortly after the release of *Pure Mathematics for Beginners,* I began receiving messages from a wide range of readers that were enjoying the book. These readers include:

- students struggling in their advanced math and computer science classes,
- physicists that wanted to learn some theoretical mathematics,
- high school math teachers that wanted to introduce their more advanced math students to mathematical theory,
- and many others.

However, I quickly learned that the book was too difficult for some readers. I refer to these readers as "pre-beginners." By a "pre-beginner," I mean a student that is ready to start learning some more advanced mathematics, but is not quite ready to dive into proofwriting.

The book you are currently reading was written with these "pre-beginners" in mind. It provides a more basic **nonrigorous** introduction to pure mathematics, while exposing readers to a wide range of mathematical topics in logic, set theory, abstract algebra, number theory, real analysis, topology, complex analysis, and linear algebra.

There are no prerequisites for this book. The content is completely self-contained. Furthermore, reading this book will naturally increase a student's level of "mathematical maturity." Although there is no single agreed upon definition of mathematical maturity, one reasonable way to define it is as "one's ability to analyze, understand, and communicate mathematics." A student with a very high level of mathematical maturity may find this book very easy—this student may want to go through the book quickly and then move on to *Pure Mathematics for Beginners*. A student with a lower level of mathematical maturity will probably find this book more challenging. However, the reward will certainly be more than worth the effort.

If you read this book and complete the exercises along the way, then your level of mathematical maturity will continually be increasing. This increased level of mathematical maturity will not only help you to succeed in advanced math courses, but it will improve your general problem solving and reasoning skills. This will make it easier to improve your performance in college, in your professional life, and on standardized tests such as the SAT, ACT, GRE, and GMAT.

At the end of each lesson there is a Problem Set. The problems in each of these Problem Sets have been organized into five levels of difficulty, followed by several Challenge Problems. Level 1 problems are the easiest and Level 5 problems are the most difficult, except for the Challenge Problems. If you want to get just a small taste of pure mathematics, then you can work on the easier problems. If you want to achieve a deeper understanding of the material, take some time to struggle with the harder problems.

The author welcomes all feedback. Feel free to email Dr. Steve Warner with any questions and comments at steve@SATPrepGet800.com.

LESSON 1
LOGIC

Statements

In mathematics, a **statement** (or **proposition**) is a sentence that can be true or false, but not both simultaneously.

Example 1.1: "Jennifer is working" is a statement because at any given time either Jennifer is working or Jennifer is not working.

Example 1.2: The sentence "Go away!" is *not* a statement because it cannot be true or false. This sentence is a **command**.

Exercise 1.3: Determine if each of the following sentences are statements:

1. The check is in the mail. _____
2. Are you feeling okay? _____
3. Unicorns are real. _____
4. Don't count your chickens before they hatch. _____
5. Odin is chasing a mouse. _____

An **atomic statement** expresses a single idea. The statement "Jennifer is working" that we discussed above is an example of an atomic statement. Let's look at a few more examples.

Example 1.4: The following sentences are atomic statements:

1. 26 is an even number.
2. Mary Tudor was the first queen of England.
3. $8 < 2$.
4. Gregory is 1.8 meters tall.
5. Life exists on planets other than the Earth.

Notes: Sentences 1 and 2 above are true atomic statements and sentence 3 is a false atomic statement.

We can't say for certain whether sentence 4 is true or false without knowing who Gregory is. However, it is either true or false. Therefore, it is a statement. Since it expresses a single idea, it is an atomic statement.

It is also unknown whether sentence 5 is true or false, but this does not change the fact that it must be either true or false. Furthermore, it expresses a single idea. Therefore, it is an atomic statement.

We use **logical connectives** to form **compound statements**. The most commonly used logical connectives are "and," "or," "if...then," "if and only if," and "not."

Example 1.5: The following sentences are compound statements:

1. 26 is an even number and $0 = 1$.

2. Ahmed is holding a piece of chalk or soda is a beverage.

3. If Odin is a cat, then elephants can breathe underwater.

4. Abraham Lincoln is alive today if and only if $5 - 3 = 2$.

5. 11 is not a prime number.

Sentence 1 above uses the logical connective "and." Since the statement "$0 = 1$" is false, it follows that sentence 1 is false. It does not matter that the statement "26 is an even number" is true. In fact, "T and F" is always F.

Sentence 2 uses the logical connective "or." Since the statement "soda is a beverage" is true, it follows that sentence 2 is true. It does not even matter whether Ahmed is holding a piece of chalk. In fact, "T or T" is always true and "F or T" is always T.

It's worth pausing for a moment to note that in the English language the word "or" has two possible meanings. There is an "inclusive or" and an "exclusive or." The "inclusive or" is true when both statements are true, whereas the "exclusive or" is false when both statements are true. In mathematics, by default, we always use the "inclusive or" unless we are told to do otherwise. To some extent, this is an arbitrary choice that mathematicians have agreed upon. However, it can be argued that it is the better choice since it is used more often and it is easier to work with. Note that we were assuming use of the "inclusive or" in the last paragraph when we said, "In fact, "T or T" is always true." See Problems 25 and 26 below for more on the "exclusive or."

Sentence 3 uses the logical connective "if...then." The statement "elephants can breathe underwater" is false. We need to know whether Odin is a cat in order to figure out the truth value of sentence 3. If Odin is a cat, then sentence 3 is false ("if T, then F" is always F). If Odin is not a cat, then sentence 3 is true ("if F, then F" is always T). Do not worry if you are confused about where the truth values just mentioned come from. We will discuss the logical connective "if...then" (as well as all the other connectives) in much more detail in the section on logical connectives below.

Sentence 4 uses the logical connective "if and only if." Since the two atomic statements have different truth values, it follows that sentence 4 is false. In fact, "F if and only if T" is always F.

Sentence 5 uses the logical connective "not." Since the statement "11 is a prime number" is true, it follows that sentence 5 is false. In fact, "not T" is always F.

Notes: (1) The logical connectives "and," "or," "if...then," and "if and only if," are called **binary connectives** because they join two statements (the prefix "bi" means "two").

(2) The logical connective "not" is called a **unary connective** because it is applied to just a single statement ("unary" means "acting on a single element").

(3) Don't worry if the meaning of any of these logical connectives confuses you. We will learn more about them in the section on logical connectives below.

Exercise 1.6: Determine if each of the following statements is an atomic statement or a compound statement.

1. Silence makes me angry. _____

2. We believe in this or we believe in that. _____

3. Charles Darwin did not believe in evolution. _____

4. My girlfriend watches the television show Rick and Morty. _____

5. If dragons are real, then I am a dragon. _____

6. Sentences that begin with the word "why" are questions. _____

7. You must find your keys or you will not be on time. _____

8. The universe will eventually collapse if and only if it is currently expanding. _____

9. The word "and" has the same meaning as the word "or." _____

10. John likes to walk, but he doesn't like to run. _____

Example 1.7: The following sentences are *not* statements:

1. When will you be back?

2. Leave me alone!

3. $x + 1 = 3$

4. This sentence is false.

5. This sentence is true.

Sentence 1 above is a question and sentence 2 is a command.

Sentence 3 has an unknown variable – it can be turned into a statement by assigning a value to the variable.

Sentences 4 and 5 are self-referential (they refer to themselves). They can be neither true nor false. Sentence 4 is called the Liar's paradox and sentence 5 is called a vacuous affirmation.

Truth Assignments

We will use letters such as p, q, r, and s to denote atomic statements. We will refer to these letters as **propositional variables**, and we will generally assign a truth value of T (for true) or F (for false) to each propositional variable. Formally, we define a **truth assignment** of a list of propositional variables to be a choice of T or F for each propositional variable in the list.

Example 1.8: Consider the propositional variable p. There are **two** possible truth assignments for this propositional variable as follows:

1. We can assign p to be true.

2. We can assign p to be false.

We can visualize this list of truth assignments with the following table:

p
T
F

Observe how the table has just one column because there is only one propositional variable. We label the column with the propositional variable p. Underneath the propositional variable, we have two rows—one for each of the two possible truth assignments.

Example 1.9: Consider the propositional variables p and q (where p and q are **distinct**). There are **four** possible truth assignments for this list of propositional variables as follows:

1. We can assign both p and q to be true.
2. We can assign p to be true and q to be false.
3. We can assign p to be false and q to be true.
4. We can assign both p and q to be false.

Note: When we say that p and q are **distinct**, we mean that $p \neq q$. In other words, we use the word distinct when we want to make sure that it is understood that the objects under consideration are different from each other.

We can visualize this list of truth assignments with the following table:

p	q
T	T
T	F
F	T
F	F

Observe how the table has two columns because there are two propositional variables. We label the columns with the propositional variables p and q. Underneath the propositional variables, we have four rows—one for each of the four possible truth assignments.

Exercise 1.10: Consider the three distinct propositional variables p, q, and r. How many different truth assignments are there for this list of propositional variables? _____

Draw a table that will allow us to visualize this list of truth assignments.

Logical Connectives

We use the symbols $\wedge$, $\vee$, $\rightarrow$, $\leftrightarrow$, and $\neg$ for the most common logical connectives. The truth value of a compound statement is determined by the truth values of its atomic parts together with applying various rules for the connectives. Let's look at each connective in detail.

We will use the "wedge" symbol $\wedge$ to represent the logical connective "and." The compound statement $p \wedge q$ is called the **conjunction** of p and q. It is pronounced "**p and q**." $p \wedge q$ is true when both p and q are true, and it is false otherwise.

The following table summarizes the truth values of $p \wedge q$ for each possible truth assignment of the propositional variables p and q.

p	q	$p \wedge q$
T	T	T
T	F	F
F	T	F
F	F	F

Notes: (1) The table displayed above is called a **truth table**. This type of table is used to display the possible truth values of a compound statement. We start by labelling the columns of the table with the propositional variables that appear in the statement, followed by the statement itself. We then use the rows to run through every possible combination of truth values for the propositional variables (all possible truth assignments) followed by the resulting truth values for the compound statement.

(2) The first two columns of the truth table above (labeled p and q) give the four possible truth assignments for the propositional variables p and q (see Example 1.9).

(3) The truth table for the conjunction is based upon the way we use the word "and" in everyday English. For example, suppose that Jamie is a girl with black hair. Then the statement "Jamie is a girl and Jamie has black hair" is true because each of the statements "Jamie is a girl" and "Jamie has black hair" are true. Similarly, the statement "Jamie is a girl and Jamie has red hair" is false because the statement "Jamie has red hair" is false. Based upon your own experience of the English language, you may be able to compute the truth value of a conjunction of two statements without needing to look back at the truth table.

Example 1.11: If p is true and q is false, then we can compute the truth value of $p \wedge q$ by looking at the second row of the truth table for the conjunction.

p	q	$p \wedge q$
T	T	T
T	F	F
F	T	F
F	F	F

We see from the highlighted row in the truth table above that $p \wedge q \equiv \text{T} \wedge \text{F} \equiv \textbf{F}$.

Note: Here the symbol ≡ can be read "is logically equivalent to." So, we see that if p is true and q is false, then $p \wedge q$ is logically equivalent to F, or more simply, $p \wedge q$ is false.

Exercise 1.12: Determine the truth value of $p \wedge q$ given that

1. p and q are both true. ___

2. p and q are both false. ___

3. p is false and q is true. ___

We will use the "vee" symbol ∨ to represent the logical connective "or." The compound statement $p \vee q$ is called the **disjunction** of p and q. It is pronounced "**p or q**." $p \vee q$ is true when p or q (or both) are true, and it is false when p and q are both false.

The following table summarizes the truth values of $p \vee q$ for each possible truth assignment of the propositional variables p and q.

p	q	$p \vee q$
T	T	T
T	F	T
F	T	T
F	F	F

Notes: (1) The truth table for the disjunction is based upon the way we use the "inclusive or" in everyday English. For example, suppose that in order to be able to watch television, Kelly's parents tell her that she must first either do the dishes or clean her room. In this context, the statement "Kelly does the dishes or Kelly cleans her room" is understood to be true if Kelly does the dishes or Kelly cleans her room **or both**. Certainly if she completes both of these tasks, it would be unnatural to penalize her.

(2) In mathematics, when we use the word "or" we always assume that we mean the "inclusive or" unless we are told otherwise. In English when we use the word "or," we are more likely to be using the "exclusive or." For example, if a waiter says "you can have fries or a salad with your order," it is unlikely that he means you can have both. Indeed, this is an example of the "exclusive or." See Problems 25 and 26 below for more on the exclusive or.

Exercise 1.13: Determine the truth value of $p \vee q$ given that

1. p and q are both true. ___

2. p and q are both false. ___

3. p is true and q is false. ___

4. p is false and q is true. ___

We will use the "rightarrow" symbol → to represent the logical connective "if...then." The compound statement $p \rightarrow q$ is called a **conditional** or **implication**. It is pronounced "**if p, then q**" or p **implies** q. $p \rightarrow q$ is true when p is false or q is true (or both), and it is false when p is true and q is false.

The following table summarizes the truth values of $p \rightarrow q$ for each possible truth assignment of the propositional variables p and q.

p	q	$p \rightarrow q$
T	T	T
T	F	F
F	T	T
F	F	T

Notes: (1) In the conditional $p \rightarrow q$, p is called the **hypothesis** (or **assumption** or **premise**) and q is called the **conclusion**.

(2) The truth table for the conditional is loosely based upon the English meaning of "if…then" or "implies." However, trying to understand this truth table by analyzing English sentences often leads to confusion. So, in this case, it may be more instructive to understand what we wish to accomplish with this connective mathematically. When should $p \rightarrow q$ be true. Well, we would like $p \rightarrow q$ to be true if the assumption that p is true forces q to be true as well (equivalently, if the hypothesis is true, then the conclusion must be true). For example, let's take the statement "If Odin is a cat, then Odin can bark." Now, if Odin happens to be a cat, then the statement just given in quotes is false. Do you see why? The hypothesis "Odin is a cat" is true. If the conditional were true, then the conclusion "Odin can bark" would be forced to be true. However, cats can't bark. So, the conditional is false. This situation corresponds to the second row in the truth table for the conditional above.

On the other hand, the statement "If Odin is a cat, then Odin can meow" is most likely true. This time the hypothesis and conclusion are both true. This situation corresponds to the first row in the truth table for the conditional above.

(3) What about the situation in which the hypothesis is false. In this case, we don't really care what the conclusion is. The conditional is true either way. When the hypothesis is false, we will say that the conditional statement is **vacuously true**. The word "vacuous" means "empty." The idea is that something that is vacuously true is true for a very silly reason. For example, suppose that we are looking at an empty room and someone says, "If there is a pig in that room, then it can fly." This is equivalent to saying, "Every pig in that room can fly." This statement is true, but for a very dumb reason. Yes, every pig in that room can fly simply because there are no pigs in the room. If there were even a single pig in the room, then the statement would be false. In order for someone to dispute our claim that every pig in the room can fly, they would need to show us a pig in the room that cannot fly. Of course, they cannot do this. After all, there are no pigs in the room. This notion of "vacuous truth" corresponds to the third and fourth rows in the truth table for the conditional above.

Exercise 1.14: Determine the truth value of $p \rightarrow q$ given that

1. p and q are both true. ___
2. p and q are both false. ___
3. p is true and q is false. ___
4. p is false and q is true. ___

We will use the "doublearrow" symbol ↔ to represent the logical connective "if and only if." The compound statement $p \leftrightarrow q$ is called a **biconditional**. It is pronounced "**p if and only if q**." $p \leftrightarrow q$ is true when p and q have the same truth value (both true or both false), and it is false when p and q have opposite truth values (one true and the other false).

The following table summarizes the truth values of $p \leftrightarrow q$ for each possible truth assignment of the propositional variables p and q.

p	q	$p \leftrightarrow q$
T	T	T
T	F	F
F	T	F
F	F	T

Exercise 1.15: Determine the truth value of $p \leftrightarrow q$ given that

1. p and q are both true. ____

2. p and q are both false. ____

3. p is true and q is false. ____

4. p is false and q is true. ____

We will use the "taildash" symbol ¬ to represent the logical connective "not." The compound statement $\neg p$ is called the **negation** of p. It is pronounced "**not p**." $\neg p$ is true when p is false, and it is false when p is true (p and $\neg p$ have opposite truth values.)

The following table summarizes the truth values of $\neg p$ for each possible truth assignment of the propositional variable p.

p	$\neg p$
T	F
F	T

Notes: (1) Since negation requires only a single propositional variable, there are just two possible truth assignments to worry about. The first column of the truth table above (labeled p) gives the two possible truth assignments (T and F).

(2) The truth table for the negation is based upon the way we use the word "not" in everyday English. For example, since the statement "Fish swim" is true, it follows that the statement "Fish do not swim" is false. Similarly, since the statement "Elephants fly" is false, it follows that the statement "Elephants do not fly" is true.

Example 1.16: If p is true, then we can compute the truth value of $\neg p$ by looking at the first row of the truth table for the negation.

p	$\neg p$
T	**F**
F	**T**

We see from the highlighted row in the truth table above that $\neg p \equiv \neg T \equiv F$.

Exercise 1.17: Determine the truth value of $\neg p$ given that p is false. ___

Example 1.18: Let p represent the statement "Ducks quack" and let q represent the statement "$0 = 1$." Note that p is true and q is false.

1. $p \wedge q$ represents "Ducks quack and $0 = 1$." Since q is false, it follows that $p \wedge q$ is false.

2. $p \vee q$ represents "Ducks quack or $0 = 1$." Since p is true, it follows that $p \vee q$ is true.

3. $p \rightarrow q$ represents "If ducks quack, then $0 = 1$." Since p is true and q is false, $p \rightarrow q$ is false.

4. $p \leftrightarrow q$ represents "Ducks quack if and only if $0 = 1$." Since p is true and q is false, $p \leftrightarrow q$ is false.

5. $\neg q$ represents the statement "$0 \neq 1$." Since q is false, $\neg q$ is true.

6. $\neg p \vee q$ represents the statement "Ducks don't quack or $0 = 1$." Since $\neg p$ and q are both false, $\neg p \vee q$ is false. Note that $\neg p \vee q$ always means $(\neg p) \vee q$. In general, without parentheses present, we always apply negation before any of the other connectives.

7. $\neg(p \vee q)$ represents the statement "It is not the case that either ducks quack or $0 = 1$." This can also be stated as "Neither do ducks quack nor is 0 equal to 1." Since $p \vee q$ is true (see 2 above), $\neg(p \vee q)$ is false.

8. $\neg p \wedge \neg q$ represents the statement "Ducks don't quack and $0 \neq 1$." This statement can also be stated as "Neither do ducks quack nor is 0 equal to 1." Since this is the same statement as in 7 above, it should follow that $\neg p \wedge \neg q$ is equivalent to $\neg(p \vee q)$. You will be asked to verify this later (see Exercise 1.27 below). For now, let's observe that since $\neg p$ is false, it follows that $\neg p \wedge \neg q$ is false. This agrees with the truth value we got in 7.

Note: The equivalence of $\neg(p \vee q)$ with $\neg p \wedge \neg q$ (see parts 7 and 8 of Example 1.18 above) is one of **De Morgan's laws**. These laws will be explored further below (see Example 1.26 and Exercise 1.27).

Exercise 1.19: Let p represent the statement "Frogs are birds," and let q represent the statement "$2 < 1$." Translate each of the following compound statements into English and compute the truth value of each statement.

1. $p \rightarrow q$ _____

2. $\neg p \vee q$ _____

3. $p \leftrightarrow q$ _____

4. $(p \rightarrow q) \wedge (q \rightarrow p)$ _____

5. $\neg(p \wedge q)$ _____

6. $\neg p \vee \neg q$ _____

Evaluating Truth

Example 1.20: Let p, q, and r be propositional variables with p and q true, and r false. Let's compute the truth value of $\neg p \vee (\neg q \rightarrow r)$.

Truth table solution: One foolproof way to compute the desired truth value is to build the whole truth table of $\neg p \vee (\neg q \rightarrow r)$ one column at a time. Since there are 3 propositional variables (p, q, and r), we will need 8 rows to get all the possible truth values (see Exercise 1.10 and its solution). We then create a column for each compound statement that appears within the given statement starting with the statements of smallest length and working our way up to the given statement. We will need columns for p, q, r (the atomic statements), $\neg p$, $\neg q$, $\neg q \rightarrow r$, and finally, the statement itself, $\neg p \vee (\neg q \rightarrow r)$. Below is the final truth table with the relevant row highlighted and the final answer circled.

p	q	r	$\neg p$	$\neg q$	$\neg q \rightarrow r$	$\neg p \vee (\neg q \rightarrow r)$
T	T	T	F	F	T	T
T	T	F	F	F	T	T
T	F	T	F	T	T	T
T	F	F	F	T	F	F
F	T	T	T	F	T	T
F	T	F	T	F	T	T
F	F	T	T	T	T	T
F	F	F	T	T	F	T

Notes: (1) We fill out the first three columns of the truth table by listing all possible combinations of truth assignments for the propositional variables p, q, and r. Notice how down the first column we have 4 T's followed by 4 F's, down the second column we alternate sequences of 2 T's with 2 F's, and down the third column we alternate T's with F's one at a time. This is a nice systematic way to make sure we get all possible combinations of truth assignments.

If you're having trouble seeing the pattern of T's and F's, here is another way to think about it: In the first column, the first half of the rows have a T and the remainder have an F. This gives 4 T's followed by 4 F's.

For the second column, we take half the number of consecutive T's in the first column (half of 4 is 2) and then we alternate between 2 T's and 2 F's until we fill out the column.

For the third column, we take half the number of consecutive T's in the second column (half of 2 is 1) and then we alternate between 1 T and 1 F until we fill out the column.

(2) Since the connective $\neg$ has the effect of taking the opposite truth value, we generate the entries in the fourth column by taking the opposite of each truth value in the first column. Similarly, we generate the entries in the fifth column by taking the opposite of each truth value in the second column.

(3) For the sixth column, we apply the connective $\rightarrow$ to the fifth and third columns, respectively, and finally, for the last column, we apply the connective $\vee$ to the fourth and sixth columns, respectively.

(4) The original question is asking us to compute the truth value of $\neg p \vee (\neg q \to r)$ when p and q are true, and r is false. In terms of the truth table, we are being asked for the entry in the second row and last (seventh) column. Therefore, the answer is **T**.

(5) This is certainly not the most efficient way to answer the given question. However, building truth tables is not too difficult, and it's a foolproof way to determine truth values of compound statements.

Alternate Solution: We have $\neg p \vee (\neg q \to r) \equiv \neg T \vee (\neg T \to F) \equiv F \vee (F \to F) \equiv F \vee T \equiv$ **T**.

Notes: (1) For the first equivalence, we simply replaced the propositional variables by their given truth values. We replaced p and q by T, and we replaced r by F.

(2) For the second equivalence, we used the first row of the truth table for the negation (drawn to the right for your convenience).

p	$\neg p$
T	F
F	T

We see from the highlighted row that $\neg T \equiv F$. We applied this result twice.

(3) For the third equivalence, we used the fourth row of the truth table for the conditional.

p	q	$p \to q$
T	T	T
T	F	F
F	T	T
F	F	T

We see from the highlighted row that $F \to F \equiv T$.

(4) For the last equivalence, we used the third row of the truth table for the disjunction.

p	q	$p \vee q$
T	T	T
T	F	T
F	T	T
F	F	F

We see from the highlighted row that $F \vee T \equiv T$.

(5) We can save a little time by immediately replacing the negation of a propositional variable by its truth value (which will be the opposite truth value of the propositional variable). For example, since p has truth value T, we can replace $\neg p$ by F. Similarly, since q has truth value T, we can replace $\neg q$ by F. The faster solution would look like this:

$$\neg p \vee (\neg q \to r) \equiv F \vee (F \to F) \equiv F \vee T \equiv \textbf{T}.$$

Quicker solution: Since q has truth value T, it follows that $\neg q$ has truth value F. So, $\neg q \to r$ has truth value T. Finally, $\neg p \vee (\neg q \to r)$ must then have truth value T.

Notes: (1) Symbolically, we can write the following:

$$\neg p \vee (\neg q \to r) \equiv \neg p \vee (\neg T \to r) \equiv \neg p \vee (F \to r) \equiv \neg p \vee T \equiv \mathbf{T}$$

(2) We can display this reasoning visually as follows:

$$\neg p \vee (\neg q \to r)$$

$$\begin{array}{c} \quad \mid T \quad \\ \quad F \quad \\ \quad T \\ \mathbf{T} \end{array}$$

The vertical lines have just been included to make sure you see which connective each truth value is written below.

We began by placing a T under the propositional variable q to indicate that q is true. Since $\neg T \equiv F$, we then place an F under the negation symbol. Next, since $F \to r \equiv T$ regardless of the truth value of r, we place a T under the conditional symbol. Finally, since $\neg p \vee T \equiv T$ regardless of the truth value of p, we place a T under the disjunction symbol. We made this last T bold to indicate that we are finished.

(3) Knowing that q has truth value T is enough to determine the truth value of $\neg p \vee (\neg q \to r)$, as we saw in Note 1 above. It's okay if you didn't notice that right away. This kind of reasoning takes a bit of practice and experience.

Exercise 1.21: Let p, q, and r be propositional variables.

1. Draw the truth table for $p \leftrightarrow (q \wedge \neg r)$.

2. Use the truth table from part 1 to compute the truth value of $p \leftrightarrow (q \wedge \neg r)$ when p is true, and q and r are false. _____

3. Suppose that p and r are both true. Is this enough information to compute the truth value of $p \leftrightarrow (q \wedge \neg r)$? _____ If so, what is that truth value? _____

19

Logical Equivalence

We say that two statements are **logically equivalent** if they have the same truth table. We use the symbol "≡" to indicate logical equivalence.

Example 1.22: Let p be a propositional variable. Let's show that p and $\neg(\neg p)$ are logically equivalent (symbolically, we write $p \equiv \neg(\neg p)$). We will show that p and $\neg(\neg p)$ have the same truth table. We can put all the information into a single table.

p	$\neg p$	$\neg(\neg p)$
T	F	T
F	T	F

Observe that the first column gives the truth values for p, the third column gives the truth values for $\neg(\neg p)$, and both these columns are identical. It follows that $p \equiv \neg(\neg p)$.

Notes: (1) The logical equivalence $p \equiv \neg(\neg p)$ is called the **law of double negation**. In words, this law says that if you negate a propositional variable twice, the resulting statement is logically equivalent to the original propositional variable.

(2) As an example in English, let p be the statement "John is hungry." Then the statement $\neg(\neg p)$ can be expressed in English as "It is not the case that John is not hungry." By the law of double negation, these two statements are logically equivalent.

Exercise 1.23: The **law of the conditional** is the logical equivalence $p \rightarrow q \equiv \neg p \lor q$. Use a truth table to verify this logical equivalence.

Note: The law of the conditional allows us to replace the conditional statement $p \rightarrow q$ by the more intuitive statement $\neg p \lor q$. Remember that we can think of the conditional statement $p \rightarrow q$ as having the hypothesis p and the conclusion q. The disjunctive form $\neg p \lor q$ tells us quite explicitly that a conditional statement is true precisely if the hypothesis p is false or the conclusion q is true (or both).

Consider the conditional statement $p \rightarrow q$. There are three other statements associated with this statement.

1. The **converse** is the statement $q \rightarrow p$.

2. The **inverse** is the statement $\neg p \rightarrow \neg q$.

3. The **contrapositive** is the statement $\neg q \rightarrow \neg p$.

Example 1.24: Consider the conditional statement "If you are a cat, then you are a mammal."

1. The converse is the statement "If you are a mammal, then you are a cat."

2. The inverse is the statement "If you are not a cat, then you are not a mammal."

3. The contrapositive is the statement "If you are not a mammal, then you are not a cat."

Notes: (1) If we let p be the statement "You are a cat" and we let q be the statement "You are a mammal," then the given conditional statement can be represented by $p \to q$. Similarly, the converse can be represented by $q \to p$, the inverse can be represented by $\neg p \to \neg q$, and the contrapositive can be represented by $\neg q \to \neg p$.

(2) Notice that in this example, the given conditional statement is true. Indeed, every cat is a mammal. On the other hand, the converse is false. After all, there are certainly mammals that are not cats. For example, a dog is a mammal that is not a cat. This example shows that a conditional statement is **not** logically equivalent to its converse.

(3) In this example, the inverse is also false. For example, a dog is not a cat, and yet, a dog is a mammal. This example shows that a conditional statement is **not** logically equivalent to its inverse.

(4) In this example, the contrapositive is true. Anything that is not a mammal cannot possibly be a cat. In fact, it turns out that a conditional statement is **always** logically equivalent to its contrapositive. We will see this in Example 1.25 below. A word of caution is in order here. The one example just given does **not** prove that the given conditional statement is logically equivalent to its contrapositive. To verify logical equivalence, we need to check the whole truth table.

Exercise 1.25: The **law of the contrapositive** is the logical equivalence $p \to q \equiv \neg q \to \neg p$. Use a truth table to verify this logical equivalence.

The **De Morgan's laws** provide formulas for negating a conjunction and for negating a disjunction.

$$\neg(p \land q) \equiv \neg p \lor \neg q \qquad \neg(p \lor q) \equiv \neg p \land \neg q$$

Example 1.26: Let's verify the first De Morgan's law. In other words, we will show that $\neg(p \land q)$ and $\neg p \lor \neg q$ are logically equivalent. We will provide two different methods.

Direct method: If $p \equiv F$ or $q \equiv F$, then $\neg(p \land q) \equiv \neg F \equiv T$ and $\neg p \lor \neg q \equiv T$ (because $\neg p \equiv T$ or $\neg q \equiv T$). If $p \equiv T$ and $q \equiv T$, then $\neg(p \land q) \equiv \neg T \equiv F$ and $\neg p \lor \neg q \equiv F \lor F \equiv F$. So, all four possible truth assignments of p and q lead to the same truth value for $\neg(p \land q)$ and $\neg p \lor \neg q$. It follows that $\neg(p \land q) \equiv \neg p \lor \neg q$.

Truth table method:

p	q	$\neg p$	$\neg q$	$p \wedge q$	$\neg(p \wedge q)$	$\neg p \vee \neg q$
T	T	F	F	T	F	F
T	F	F	T	F	T	T
F	T	T	F	F	T	T
F	F	T	T	F	T	T

Observe that the sixth column gives the truth values for $\neg(p \wedge q)$, the seventh column gives the truth values for $\neg p \vee \neg q$, and both these columns are identical. It follows that $\neg(p \wedge q) \equiv \neg p \vee \neg q$.

Exercise 1.27: Use a truth table to verify the second De Morgan's law $\neg(p \vee q) \equiv \neg p \wedge \neg q$.

List 1.28: Here is a list of some useful logical equivalences. The dedicated reader may want to verify each of these by drawing a truth table or by using direct arguments similar to that used in Example 1.26 (some of these are asked for in Problems 57 through 62 below).

1. **Law of double negation:** $p \equiv \neg(\neg p)$

2. **De Morgan's laws:** $\neg(p \wedge q) \equiv \neg p \vee \neg q$ $\qquad$ $\neg(p \vee q) \equiv \neg p \wedge \neg q$

3. **Commutative laws:** $\quad p \wedge q \equiv q \wedge p$ $\qquad\qquad$ $p \vee q \equiv q \vee p$

4. **Associative laws:** $\quad (p \wedge q) \wedge r \equiv p \wedge (q \wedge r)$ $\qquad$ $(p \vee q) \vee r \equiv p \vee (q \vee r)$

5. **Distributive laws:** $\quad p \wedge (q \vee r) \equiv (p \wedge q) \vee (p \wedge r)$ $\quad$ $p \vee (q \wedge r) \equiv (p \vee q) \wedge (p \vee r)$

6. **Identity laws:** $\qquad p \wedge \text{T} \equiv p \qquad p \wedge \text{F} \equiv \text{F} \qquad p \vee \text{T} \equiv \text{T} \qquad p \vee \text{F} \equiv p$

7. **Negation laws:** $\qquad p \wedge \neg p \equiv \text{F}$ $\qquad\qquad\qquad$ $p \vee \neg p \equiv \text{T}$

8. **Redundancy laws:** $\quad p \wedge p \equiv p$ $\qquad\qquad\qquad$ $p \vee p \equiv p$

9. **Absorption laws:** $\quad (p \vee q) \wedge p \equiv p$ $\qquad\qquad$ $(p \wedge q) \vee p \equiv p$

10. **Law of the conditional:** $\quad p \rightarrow q \equiv \neg p \vee q$

11. **Law of the contrapositive:** $p \rightarrow q \equiv \neg q \rightarrow \neg p$

12. **Law of the biconditional:** $\quad p \leftrightarrow q \equiv (p \rightarrow q) \wedge (q \rightarrow p)$

Note: Although this is a fairly long list of laws, a lot of it is quite intuitive. For example, in English the word "and" is commutative. The statements "I have a cat and I have a dog" and "I have a dog and I have a cat" have the same meaning. So, it's easy to see that $p \wedge q \equiv q \wedge p$ (the first law in 3 above). As another example, the statement "I have a cat and I do not have a cat" could never be true. So, it's easy to see that $p \wedge \neg p \equiv \text{F}$ (the first law in 7 above).

Example 1.29: Let's show that the statement $p \wedge [(p \wedge \neg q) \vee q]$ is logically equivalent to the atomic statement p.

Solution:
$$p \wedge [(p \wedge \neg q) \vee q] \equiv p \wedge [q \vee (p \wedge \neg q)] \equiv p \wedge [(q \vee p) \wedge (q \vee \neg q)] \equiv p \wedge [(q \vee p) \wedge \mathrm{T}]$$
$$\equiv p \wedge (q \vee p) \equiv (q \vee p) \wedge p \equiv (p \vee q) \wedge p \equiv p$$

So, we see that $p \wedge [(p \wedge \neg q) \vee q]$ is logically equivalent to the atomic statement p.

Notes: (1) For the first equivalence, we used the second commutative law.

(2) For the second equivalence, we used the second distributive law.

(3) For the third equivalence, we used the second negation law.

(4) For the fourth equivalence, we used the first identity law.

(5) For the fifth equivalence, we used the first commutative law.

(6) For the sixth equivalence, we used the second commutative law.

(7) For the last equivalence, we used the first absorption law.

Exercise 1.30: Show that the statement $[(\neg p \vee q) \wedge p] \vee q$ is logically equivalent to q.

_____
_____
_____

Tautologies and Contradictions

A statement that has truth value T for all truth assignments of the propositional variables is called a **tautology**. Similarly, a statement that has truth value F for all truth assignments of the propositional variables is called a **contradiction**.

Example 1.31: Let's show that the statement $p \rightarrow p$ is a tautology.

Direct method: If $p \equiv \mathrm{T}$, then $p \rightarrow p \equiv \mathrm{T} \rightarrow \mathrm{T} \equiv \mathrm{T}$. If $p \equiv \mathrm{F}$, then $p \rightarrow p \equiv \mathrm{F} \rightarrow \mathrm{F} \equiv \mathrm{T}$. Since both possible truth assignments of the propositional variable p lead to the statement $p \rightarrow p$ having truth value T, it follows that $p \rightarrow p$ is a tautology.

Truth table method:

p	$p \rightarrow p$
T	T
F	T

Since the last column of the truth table consists of only the truth value T, the statement $p \rightarrow p$ is a tautology.

Exercise 1.32: Use a truth table to show that $(p \rightarrow q) \leftrightarrow (\neg q \rightarrow \neg p)$ is a tautology.

Note: Observe the similarity between the tautology $(p \rightarrow q) \leftrightarrow (\neg q \rightarrow \neg p)$ and the law of the contrapositive $p \rightarrow q \equiv \neg q \rightarrow \neg p$ (this is logical equivalence 11 from List 1.28). Given any logical equivalence $A \equiv B$ (where A and B are sentences), we always have a corresponding tautology $A \leftrightarrow B$.

Example 1.33: From the first De Morgan's Law $\neg(p \wedge q) \equiv \neg p \vee \neg q$, it follows that the statement $\neg(p \wedge q) \leftrightarrow \neg p \vee \neg q$ is a tautology.

Exercise 1.34: Show directly that the statement $p \wedge \neg p$ is a contradiction (by "directly," we mean that you should **not** use a truth table).

Problem Set 1

Full solutions to these problems are available for free download here:

www.SATPrepGet800.com/PMNR2ZX

LEVEL 1

Determine whether each of the following sentences is an atomic statement, a compound statement, or not a statement at all:

1. Grace is not going shopping tomorrow.

2. Where did I go wrong?

3. Stay out of my way today.

4. $x > 26$.

5. I visited the law firm of Cooper and Smith.

6. If there is an elephant in the room, then we need to talk.

7. $2 < -7$ or $15 > 100$.

8. This sentence is six words long.

9. A triangle is equilateral if and only if all three sides of the triangle have the same length.

10. I cannot speak Russian, but I can speak Spanish.

What is the negation of each of the following statements?

11. Cauliflower is Jamie's favorite vegetable.

12. We have three cats.

13. $15 < -12$.

14. You are not serious.

15. The function f is continuous.

16. The real number system with the standard topology is locally compact.

LEVEL 2

Let p represent the statement "5 is an odd integer," let q represent the statement "Brazil is in Europe," and let r represent the statement "A lobster is an insect." Rewrite each of the following symbolic statements in words, and state the truth value of each statement:

17. $p \lor q$

18. $\neg r$

19. $p \rightarrow q$

20. $p \leftrightarrow r$

21. $\neg q \land r$

22. $\neg(p \land q)$

23. $\neg p \lor \neg q$

24. $(p \land q) \rightarrow r$

Consider the compound sentence "You can have a cookie or ice cream." In English this would most likely mean that you can have one or the other but not both. The word "or" used here is generally called an "exclusive or" because it excludes the possibility of both. The disjunction is an "inclusive or."

25. Using the symbol $\oplus$ for exclusive or, draw the truth table for this connective.

26. Using only the logical connectives $\neg$, $\land$, and $\lor$, produce a statement using the propositional variables p and q that has the same truth values as $p \oplus q$.

LEVEL 3

Consider the four distinct propositional variables p, q, r, and s.

27. How many different truth assignments are there for this list of propositional variables?

28. How many different truth assignments are there for this list of propositional variables such that p is true and q is false?

29. How many different truth assignments are there for this list of propositional variables such that q, r, and s are all true?

30. How many different truth assignments are there for a list of 5 propositional variables?

Let p, q, and r represent true statements. Compute the truth value of each of the following compound statements:

31. $(p \lor q) \lor r$

32. $(p \lor q) \land \neg r$

33. $\neg p \to (q \lor r)$

34. $\neg(p \leftrightarrow \neg q) \land r$

35. $\neg[p \land (\neg q \to r)]$

36. $\neg[(\neg p \lor \neg q) \leftrightarrow \neg r]$

37. $p \to (q \to \neg r)$

38. $\neg[\neg p \to (q \to \neg r)]$

Determine if each of the following statements is a tautology, a contradiction, or neither.

39. $p \land p$

40. $p \land \neg p$

41. $(p \lor \neg p) \to (p \land \neg p)$

42. $\neg(p \lor q) \leftrightarrow (\neg p \land \neg q)$

43. $p \to (\neg q \land r)$

44. $(p \leftrightarrow q) \to (p \to q)$

LEVEL 4

Assume that the given compound statement is true. Determine the truth value of each propositional variable.

45. $p \land q$

46. $\neg(p \to q)$

47. $p \leftrightarrow [\neg(p \land q)]$

48. $[p \land (q \lor r)] \land \neg r$

Let p represent a true statement. Decide if this is enough information to determine the truth value of each of the following statements. If so, state that truth value.

49. $p \lor q$

50. $p \to q$

51. $\neg p \to \neg(q \lor \neg r)$

52. $\neg(\neg p \land q) \leftrightarrow p$

53. $(p \leftrightarrow q) \leftrightarrow \neg p$

54. $\neg[(\neg p \land \neg q) \leftrightarrow \neg r]$

55. $[(p \land \neg p) \to p] \land (p \lor \neg p)$

56. $r \to [\neg q \to (\neg p \to \neg r)]$

For each of the following pairs of statements A and B, show that $A \equiv B$.

57. $A = p \land q$, $B = q \land p$

58. $A = (p \lor q) \lor r$, $B = p \lor (q \lor r)$

59. $A = p \land (q \lor r)$, $B = (p \land q) \lor (p \land r)$

60. $A = (p \lor q) \land p$, $B = p$

61. $A = p \leftrightarrow q$, $B = (p \to q) \land (q \to p)$

62. $A = \neg(p \to q)$, $B = p \land \neg q$

Simplify each statement.

63. $p \lor (p \land \neg p)$

64. $(p \land q) \lor \neg p$

65. $\neg p \to (\neg q \to p)$

66. $(p \land \neg q) \lor p$

67. $[(q \land p) \lor q] \land [(q \lor p) \land p]$

Without drawing a truth table or using List 1.28, show that each of the following is a tautology.

68. $[p \wedge (q \vee r)] \leftrightarrow [(p \wedge q) \vee (p \wedge r)]$

69. $[[(p \wedge q) \to r] \to s] \to [(p \to r) \to s]$

Let n be a positive integer (in other words, n is one of the numbers 1, 2, 3, 4, ...) and let A be a statement involving n propositional variables. Determine how many rows are in the truth table for A if n is equal to each of the following:

70. $n = 6$

71. $n = 10$

72. n is an arbitrary positive integer (provide an explicit expression involving n.)

CHALLENGE PROBLEMS

73. Determine a tautology or contradiction containing *at least* three propositional variables and *at least* three logical connectives so that the truth values for all truth assignments can be evaluated with *no more than* 5 computations and such that *at least* 3 computations are required. Write out your compound statement, followed by your 3 to 5 computations.

74. Let T be a truth table. Explain why there is a statement A involving only the logical connectives $\wedge$, $\vee$, and $\neg$ such that the truth table of A is T.

    Hint: For example, let T be the following truth table:

p	q	?
T	T	T
T	F	F
F	T	T
F	F	F

    If we let A be the statement $(p \wedge q) \vee (\neg p \wedge q)$, then the truth table of A is T.

LESSON 2
SET THEORY

Describing Sets Explicitly

A **set** is simply a collection of "objects." These objects can be

- numbers such as 1, 2, or 3.

- letters such as a, b, t, or x.

- shapes such as △, □, or ◯.

- animals such as bear, dolphin, starfish, or toad.

- ... or just about anything else you can imagine.

We will usually refer to the objects in a set as the **members** or **elements** of the set.

If a set consists of a small number of elements, we can describe the set simply by listing the elements in the set inside curly braces, separating elements by commas. We call this method of describing a set the **roster method**.

Example 2.1:

1. {zebra, hippopotamus} is the set consisting of two elements: *zebra* and *hippopotamus*.

2. {gargoyle, fly, teacher, motorcycle, chainsaw, hydrogen} is the set consisting of six elements: *gargoyle*, *fly*, *teacher*, *motorcycle*, *chainsaw*, and *hydrogen*.

3. {0, 1, 2, 3, 4} is the set consisting of five elements: 0, 1, 2, 3, and 4. The elements in this set happen to be *numbers*.

Exercise 2.2: Determine how many elements are in each of the following sets and then list the elements in the set.

1. $\{p, q, r, s, t\}$ _____

2. {centipede, pencil, artichoke} _____

3. {2.5, 7.01, 11.3, 19.65} _____

4. {blue, red, yellow, green, purple, orange} _____

5. {Zeus, Hera, Andromeda, Hermes} _____

A set is determined by its elements and not the order in which the elements are presented. For example, the set {2, 0, 3, 1, 4} is the same as the set {0, 1, 2, 3, 4}.

Also, the set {0, 0, 1, 2, 2, 2, 3, 3, 4} is the same as the set {0, 1, 2, 3, 4}. If we are describing a set by listing its elements, the most natural way to do so is to list each element just once.

We will usually name sets using capital letters such as A, B, C,..., and so on. For example, we might write $A = \{a, b, c, d\}$. So, A is the set consisting of the elements a, b, c, and d.

Example 2.3: Consider the sets $X = \{x, y\}$, $Y = \{y, x\}$, $Z = \{x, y, x, y, x\}$. Then X, Y, and Z all represent the same set. We can write $X = Y = Z$.

Exercise 2.4: For each of the following, circle the set that is **not** equal to the others.

1. $\{2, 4, 6\}$, $\{4, 2, 6\}$, $\{1, 1, 4, 6\}$, $\{2, 2, 4, 6, 6\}$
2. $\{a, x, t, v, d, b\}$, $\{a, a, x, x, x, t, t, t, t, v, v, v, d, d\}$, $\{t, a, x, d, v, d, x, a, t\}$, $\{t, a, x, x, v, d\}$
3. $\{\text{cat}, \text{dog}\}$, $\{c, a, t, d, o, g\}$, $\{\text{dog}, \text{cat}, \text{dog}\}$, $\{\text{cat}, \text{cat}, \text{dog}, \text{cat}\}$

We use the symbol $\in$ to indicate membership. Specifically, $x \in A$ means "x is an element of A," whereas $x \notin A$ means "x is **not** an element of A." We will often simply say "x is in A," and "x is not in A," respectively. Membership is an example of a **relation**. It describes a relationship between objects.

Example 2.5: Let $D = \{x, k, 9, \delta, \square\}$. Then $x \in D$, $k \in D$, $9 \in D$, $\delta \in D$, and $\square \in D$. We can combine all this information into a single statement as follows: $x, k, 9, \delta, \square \in D$.

Note: δ (pronounced "delta") is a greek letter. It is the fourth letter of the greek alphabet.

Exercise 2.6: Let $Y = \{0, 5, 12, z, t, \text{eagle}\}$. Determine if each of the following is true or false.

1. $5 \in Y$
2. $z \notin Y$
3. $\text{hawk} \in Y$
4. 6 is a member of Y
5. y is not an element of Y
6. $12, t \in Y$
7. $0, 5, 12, \text{eagle} \in Y$
8. $x, z, t \in Y$

Describing Sets with Ellipses

If a set consists of many elements, we can use **ellipses** (...) to help describe the set. For example, the set consisting of the natural numbers between 1 and 100, inclusive, can be written $\{1, 2, 3, ..., 99, 100\}$ ("inclusive" means that we include 1 and 100). The ellipses between 3 and 99 are there to indicate that there are elements in the set that we are not explicitly mentioning.

Example 2.7: Let $K = \{0, 2, 4, ..., 20\}$. Then we have $6 \in K$ and $18 \in K$, whereas $9 \notin K$. Using the roster method, we have $K = \{0, 2, 4, 6, 8, 10, 12, 14, 16, 18, 20\}$.

Exercise 2.8: Let $L = \{1, 3, 5, ..., 25\}$. Use the roster method to describe the set L.

We can also use ellipses to help describe **infinite sets**. The set of **natural numbers** can be written $\mathbb{N} = \{0, 1, 2, 3, \ldots\}$, and the set of **integers** can be written $\mathbb{Z} = \{\ldots, -4, -3, -2, -1, 0, 1, 2, 3, 4, \ldots\}$.

Notes: (1) Some mathematicians exclude 0 from the set of natural numbers. In this book, 0 will always be included. Symbolically, $0 \in \mathbb{N}$.

(2) Notice how we use special character symbols to represent the natural numbers and integers. The characters $\mathbb{N}$ and $\mathbb{Z}$ are said to be **doublestruck** or in **blackboard bold**. In general, important sets are written using doublestruck character symbols. The natural numbers and integers are two such examples. We will see several more shortly.

Example 2.9:

1. The **even natural numbers** can be written $\mathbb{E} = \{0, 2, 4, 6, \ldots\}$. We may also use the notation $2\mathbb{N}$ to describe this set. So, $2\mathbb{N} = \mathbb{E} = \{0, 2, 4, 6, \ldots\}$.

2. The **odd natural numbers** can be written $\mathbb{O} = \{1, 3, 5, \ldots\}$. We may also use the notation $2\mathbb{N} + 1$ to describe this set. So, $2\mathbb{N} + 1 = \mathbb{O} = \{1, 3, 5, \ldots\}$.

3. The **positive integers** can be written $\mathbb{Z}^+ = \{1, 2, 3, 4, \ldots\}$ or $\mathbb{N}^+ = \{1, 2, 3, 4, \ldots\}$ (the positive integers and the positive natural numbers describe the same set).

Exercise 2.10: Use brackets, commas, and ellipses to describe each of the following sets:

1. The **even integers** $2\mathbb{Z}$. _____

2. The **odd integers** $2\mathbb{Z} + 1$. _____

3. The **negative integers** $\mathbb{Z}^-$. _____

Describing Sets with Properties

Another way to describe a set is to specify a certain property P that all its elements have in common. There are endless possibilities for what a property could be. For example, suppose that P is the property of being a color. Then "red" satisfies the property P, whereas "lamp" does **not** satisfy the property P.

If we wish to describe a set with a property P, then we can use the **set-builder notation** $\{x | P(x)\}$. The expression $\{x | P(x)\}$ can be read "the set of all x such that the property $P(x)$ is true." Note that the symbol "|" is read as "such that."

The letter x in the set-builder notation is called a **variable**. The choice of x is completely arbitrary. We could have used any other letter or symbol. In other words, $\{x | P(x)\}$, $\{t | P(t)\}$, $\{\square | P(\square)\}$, and $\{? | P(?)\}$ all have exactly the same meaning. The idea here is that we substitute objects in for the variable. If that object has the given property, then it is in the set. If it does not have the given property, then it is not in the set. Sticking with our example above, let $A = \{x \mid x \text{ is a color}\}$. Then red $\in A$ ("red" is an element of the set A) because when we substitute "red" in for the variable x, we get a true statement. Indeed, red is a color. On the other hand, lamp $\notin A$ ("lamp" is **not** an element of the set A) because when we substitute "lamp" in for the variable x, we get a false statement. Indeed, the statement "lamp is a color" is false.

Example 2.11: Let $B = \{x \mid x$ is a beverage$\}$. In words, we can describe the set B as "the set of all x such that x is a beverage." Water is an element of this set because water is a beverage. In other words, if we replace x by "water," then we get the true statement "water is a beverage." Symbolically, we can write water $\in B$. Tiger is not an element of this set because a tiger is not a beverage. Symbolically, we can write tiger $\notin B$. However, we do have tiger $\in \{x \mid x$ is an animal$\}$. Similarly, we also have tiger $\in \{x \mid x$ is a cat that roars$\}$.

Exercise 2.12: Let $X = \{x \mid x$ is a word containing at least three distinct vowels$\}$. Determine if each of the following is in the set X.

1. hello ____

2. cuttlefish ____

3. gone away ____

4. existential ____

5. sediment ____

Example 2.13: Let's look at a few different ways that we can describe the set $\{0, 1, 2, 3, 4\}$. We have already seen above that reordering and/or repeating elements does not change the set. For example, $\{1, 1, 0, 4, 3, 3, 2, 2, 2\}$ describes the same set. Here are a few descriptions using set-builder notation:

- $\{n \mid n$ is a natural number between 0 and 4, inclusive$\}$

- $\{n \mid n$ is an integer between 0 and 4, inclusive$\}$

- $\{k \mid k \in \mathbb{N} \wedge 0 \leq k \leq 4\}$

- $\{t \mid t \in \mathbb{Z} \wedge 0 \leq t < 5\}$

- $\{m \mid m = 0, 1, 2, 3, \text{ or } 4\}$

Notes: (1) The first expression in the bulleted list above can be read "the set of n such that n is a natural number between 0 and 4, inclusive." Recall that the word "inclusive" means that we include 0 and 4.

(2) The second expression can be read "the set of n such that n is an integer between 0 and 4, inclusive."

(3) The third expression can be read "the set of k such that k is a natural number and k is between 0 and 4, including both 0 and 4. Note that the abbreviation "$k \in \mathbb{N}$" can be read "k is in the set of natural numbers," or more briefly, "k is a natural number." Also note that the symbol "$\wedge$" is the conjunction from Lesson 1, and it is read as "and." We used the letter "k" for the variable here (as opposed to the letter "n" that was used in the first two expressions). Once again, we can use any variable name we like, as long as it doesn't lead to confusion.

(4) The fourth expression can be read "the set of t such that t is an integer and t is between 0 and 5, including 0 and excluding 5. Note that the abbreviation "$t \in \mathbb{Z}$" can be read "t is in the set of integers," or more briefly, "t is an integer."

(5) The fifth expression can be read "the set of m such that m is 0, 1, 2, 3, or 4."

Exercise 2.14: Use set-builder notation to describe each of the following sets.

1. $\{0, 2, 4, 6, 8, 10, 12\}$ _____
2. $\{-3, -1, 1, 3, 5, \dots, 97, 99\}$ _____
3. $\mathbb{N}$ _____
4. $\mathbb{Z}$ _____
5. $2\mathbb{Z}$ _____

For easier readability, we may include a **bounding set** when using set-builder notation. If we wish to describe a set with a property P and a bounding set A, then the corresponding set-builder notation is $\{x \in A \mid P(x)\}$. As an example, consider the set $\{7, 8, 9\}$. Since every element of this set is a natural number, we can use the set of natural numbers, $\mathbb{N}$, as a bounding set. So, instead of writing $\{k \mid k \in \mathbb{N} \land 7 \le k \le 9\}$, we can use the friendlier notation $\{k \in \mathbb{N} \mid 7 \le k \le 9\}$. Notice how the bounding set appears to the **left** of the vertical line.

Example 2.15: In example 2.13, we described the set $\{0, 1, 2, 3, 4\}$ in various ways using set-builder notation. Let's look at a few more ways to do this, this time using bounding sets in the description.

* $\{n \in \mathbb{N} \mid n \text{ is between 0 and 4, inclusive}\}$
* $\{n \in \mathbb{Z} \mid n \text{ is between 0 and 4, inclusive}\}$
* $\{k \in \mathbb{N} \mid 0 \le k \le 4\}$
* $\{t \in \mathbb{Z} \mid 0 \le t < 5\}$
* $\{m \in \mathbb{N} \mid m = 0, 1, 2, 3, \text{ or } 4\}$

In addition to the sets $\mathbb{N}$ (the natural numbers) and $\mathbb{Z}$ (the integers), let's look at a few more sets that will show up throughout this book.

The set of **rational numbers** is $\mathbb{Q} = \left\{\frac{a}{b} \mid a, b \in \mathbb{Z} \text{ and } b \ne 0\right\}$. In words, $\mathbb{Q}$ is "the set of quotients a over b such that a and b are integers and b is not zero." Some examples of rational numbers are $\frac{0}{5}, \frac{2}{3}, \frac{5}{9}$, and $\frac{-6}{7}$. We identify rational numbers $\frac{a}{b}$ and $\frac{c}{d}$ whenever $ad = bc$. For example, $\frac{1}{2} = \frac{3}{6}$ because $1 \cdot 6 = 2 \cdot 3$. We also abbreviate the rational number $\frac{a}{1}$ as a. In this way, we can think of every integer as a rational number. For example, we have $\frac{0}{5} = \frac{0}{1}$ (because $0 \cdot 1 = 5 \cdot 0$), and therefore, we can abbreviate $\frac{0}{5}$ as 0. Similarly, we can abbreviate $\frac{15}{3}$ as 5 (because $\frac{15}{3} = \frac{5}{1}$).

Exercise 2.16: Place the following rational numbers into six groups so that any two rational numbers in the same group are equal, while any two rational numbers in different groups are not equal

$$\frac{3}{5} \quad \frac{-13}{7} \quad \frac{0}{6} \quad \frac{6}{10} \quad \frac{4}{4} \quad 0 \quad \frac{13}{-7} \quad \frac{-5}{-3} \quad \frac{-17}{-17} \quad 1 \quad \frac{1}{1} \quad \frac{10}{-6} \quad \frac{0}{-1} \quad \frac{-15}{9} \quad \frac{20}{12}$$

_____   _____   _____   _____   _____   _____

If a and b are integers and $b \neq 0$, then the expression $\frac{a}{b}$ is called a **fraction**. So, the set of rational numbers, $\mathbb{Q}$, can also be referred to as the set of fractions. Each fraction can also be represented in another way, namely as a **decimal**. We will discuss this a bit more after defining the real numbers.

To define the set of **real numbers**, $\mathbb{R}$, we first define a **digit** to be one of the symbols $0, 1, 2, 3, 4, 5, 6, 7, 8$, or 9. We then define $\mathbb{R}$ to be the set of numbers of the form $x.y$ (the dot between the x and y is called a **decimal point** and the number $x.y$ is called a **decimal**), where $x \in \mathbb{Z}$ and y is an infinite "string" of digits without a **tail of 9's** (meaning there are infinitely many digits in the string that are **not** 9). Symbolically, we have

$$\mathbb{R} = \{x.y \mid x \in \mathbb{Z} \text{ and } y \text{ is an infinite string of digits without a tail of 9's}\}.$$

Some examples of real numbers are $0.000\ldots$, $0.333\ldots$, $-16.000\ldots$, and $1.010010001\ldots$ We will generally delete tails of 0's. So, we would write $0.000\ldots$ as 0 and $-16.000\ldots$ as -16. We will not consider $53.023999999\ldots$ to be a real number because of the tail of 9's (an alternative approach would be to identify $53.023999999\ldots$ with 53.024). We can visualize the set of real numbers with the **real line**.

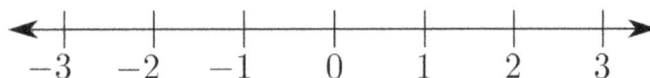

Earlier, we mentioned that we could represent every fraction (rational number) as a decimal. There are two practical ways to do this.

1. Type the fraction into a calculator and press ENTER.

2. Perform long division.

For example, we can represent the fraction $\frac{3}{2}$ as the decimal 1.5 and we can represent the fraction $\frac{2}{3}$ as the real number $0.66666\ldots$ We may abbreviate this last number by using the notation $0.\overline{6}$. The "bar" over the 6 indicates that the 6 repeats forever. As another example of this notation, the number $0.12\overline{345}$ abbreviates $0.12345345345345\ldots$ Notice how the 1 and 2 appear just once (because the bar is not over those digits), whereas the 3, 4, and 5 repeat forever.

Every rational number can be represented as a decimal that either **terminates** (has a tail of 0's) or **repeats** (has a tail with a finite repeating pattern). Any decimal (real number) that does **not** terminate or repeat is called an **irrational number**.

Exercise 2.17: Determine if each real number is a rational number or an irrational number.

1. 2.578 _____

2. $1.515151515\ldots$ _____

3. $2.02002000200002000002\ldots$ _____

4. $-1.23456789123456789123456789\ldots$ _____

5. $1.23456789101112131415161718 19\ldots$ _____

The **complex numbers** are defined as $\mathbb{C} = \{a + bi \mid a, b \in \mathbb{R}\}$. In words, $\mathbb{C}$ is "the set of $a + bi$ such that a and b are real numbers." Some examples of complex numbers are $0 + 0i$, $-2 + 0i$, $1 + 2i$, $2.3 - 5i = 2.3 + (-5)i$, and $4.235235235\ldots + 51.2020020002\ldots i$. We will abbreviate $0 + 0i$ as 0, $a + 0i$ as a, and $0 + bi$ as bi. For example, $-2 + 0i = -2$. By identifying $a + 0i$ as a, we can think of every real number as a complex number. Complex numbers of the form bi are called **pure imaginary numbers**.

If we identify $1 = 1 + 0i$ with the point $(1, 0)$, and we identify $i = 0 + 1i$ with the point $(0, 1)$, then it is natural to write the complex number $a + bi$ as the point (a, b). Here is a reasonable justification for this: $a + bi = a(1, 0) + b(0, 1) = (a, 0) + (0, b) = (a, b)$

In this way, we can visualize a complex number as a point in **The Complex Plane**. A portion of the Complex Plane is shown to the right with several complex numbers displayed as points of the form (x, y).

The Complex Plane is formed by taking two copies of the real line and placing one horizontally and the other vertically. The horizontal copy of the real line is called the x-axis or the **real axis** (labeled x in the figure) and the vertical copy of the real line is called the y-axis or **imaginary axis** (labeled y in the figure). The two axes intersect at the point $(0, 0)$. This point is called the **origin**.

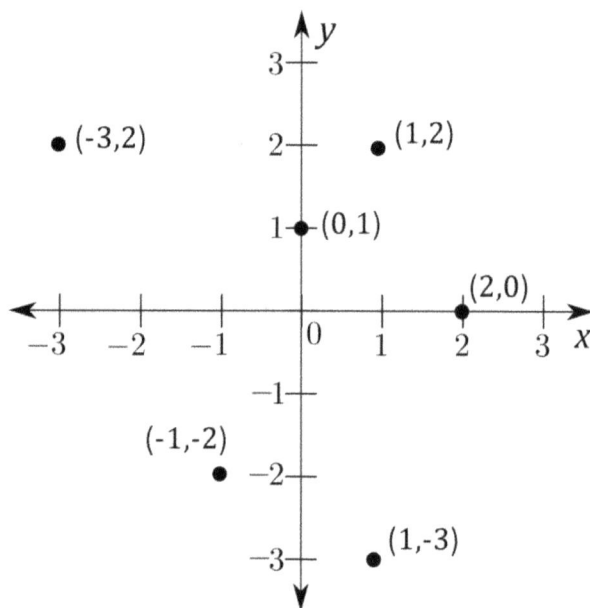

Exercise 2.18: Plot the following points in the Complex Plane:

1. 0
2. -1.5
3. $1 + i$
4. $-2 + \frac{5}{2}i$
5. $-1.\overline{3}i$

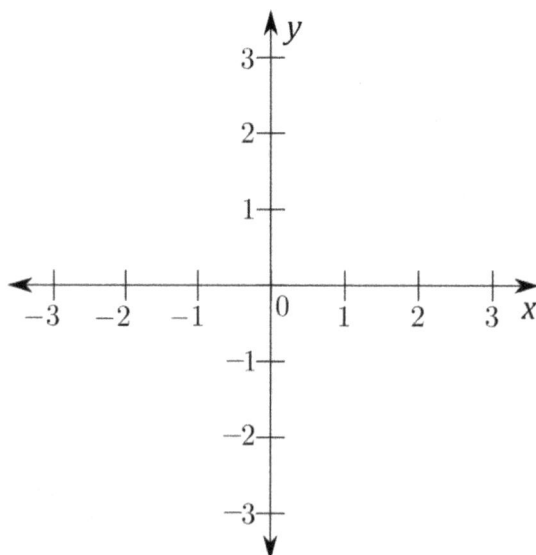

The **empty set** is the unique set with no elements. We use the symbol $\emptyset$ to denote the empty set (some authors use the symbol $\{\}$ instead).

Cardinality of a Finite Set

If A is a finite set, we define the **cardinality** of A, written $|A|$, to be the number of elements of A. For example, $|\{a, b, c\}| = 3$.

Example 2.19: Let $A = \{$cat, dog, horse, walrus$\}$, $B = \{s, t, t\}$, $C = \{25, 26, 27, \dots, 3167, 3168\}$, $D = \{\{a\}, \{a, a\}, \{a, a, a\}\}$, and $E = \emptyset$. Then $|A| = 4$, $|B| = 2$, $|C| = 3144$, $|D| = 1$, and $|E| = 0$.

Notes: (1) The set A consists of the four elements "cat," "dog," "horse," and "walrus."

(2) The set B consists of just two elements: s and t. Remember that $\{s, t, t\} = \{s, t\}$.

(3) The number of consecutive integers from m to n, inclusive, is $n - m + 1$. For set C, we have $m = 25$ and $n = 3168$. Therefore, $|C| = 3168 - 25 + 1 = 3144$.

(4) I call the formula "$n - m + 1$" the **fence-post formula**. If you construct a 3-foot fence by placing a fence-post every foot, then the fence will consist of 4 fence-posts ($3 - 0 + 1 = 4$).

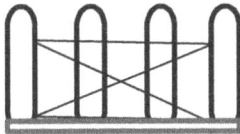

(5) Since $\{a, a\} = \{a\}$ and $\{a, a, a\} = \{a\}$, it follows that $D = \{\{a\}, \{a\}, \{a\}\} = \{\{a\}\}$. So, D consists of the single element $\{a\}$.

(6) Remember that $\emptyset$ (pronounced "the empty set") is the unique set with no elements.

Exercise 2.20: Determine the cardinality of each of the following sets:

1. $\{1, 2, 3, \dots, 50\}$ _____

2. $\{c, d, e, f, e, d, c\}$ _____

3. $\{\emptyset, \{\emptyset\}\}$ _____

4. $\{n \in \mathbb{N} \mid 126 \leq n \leq 2007\}$ _____

5. $\{x, \{x, x\}, \{x, x, x\}, \{x, \{x, x\}\}\}$ _____

Subsets and Proper Subsets

We say that a set A is a **subset** of a set B, written $A \subseteq B$, if every element of A is an element of B.

Example 2.21:

1. Let $A = \{1, 2\}$ and $B = \{1, 2, 3\}$. The only elements of A are 1 and 2. Since 1 and 2 are also elements of B, we see that $A \subseteq B$.

   Notice that $B \nsubseteq A$ (B is **not** a subset of A) because $3 \in B$, but $3 \notin A$.

2. Let $\mathbb{N} = \{0, 1, 2, 3, \dots\}$ be the set of natural numbers and let $\mathbb{Z} = \{\dots, -3, -2, -1, 0, 1, 2, 3, 4, \dots\}$ be the set of integers. Since every natural number is an integer, $\mathbb{N} \subseteq \mathbb{Z}$.

3. By making appropriate identifications, we have the following sequence of inclusions:

$$\mathbb{N} \subseteq \mathbb{Z} \subseteq \mathbb{Q} \subseteq \mathbb{R} \subseteq \mathbb{C}.$$

In general, if $A \subseteq B$ and $B \subseteq C$, then $A \subseteq C$ (we say that $\subseteq$ is **transitive**). In this way we see that we have many other inclusions such as $\mathbb{N} \subseteq \mathbb{Q}$, $\mathbb{N} \subseteq \mathbb{R}$,…, and so on.

4. Consider the sets $2\mathbb{Z} = \{…, -6, -4, -2, 0, 2, 4, 6, …\}$ and $4\mathbb{Z} = \{…, -12, -8, -4, 0, 4, 8, 12, …\}$. Then $4\mathbb{Z} \subseteq 2\mathbb{Z}$. Note that the opposite inclusion is false. That is, $2\mathbb{Z} \not\subseteq 4\mathbb{Z}$. To see this, we just need a single **counterexample** (a counterexample is an example that is used to show that a statement is false). Well, we have $2 \in 2\mathbb{Z}$, but $2 \notin 4\mathbb{Z}$.

5. Consider the sets $2\mathbb{Z} = \{…, -6, -4, -2, 0, 2, 4, 6, …\}$ and $3\mathbb{Z} = \{…, -9, -6, -3, 0, 3, 6, 9, …\}$. Neither of these sets is a subset of the other. To see that $2\mathbb{Z} \not\subseteq 3\mathbb{Z}$, observe that $2 \in 2\mathbb{Z}$, whereas $2 \notin 3\mathbb{Z}$. To see that $3\mathbb{Z} \not\subseteq 2\mathbb{Z}$, observe that $3 \in 3\mathbb{Z}$, whereas $3 \notin 2\mathbb{Z}$.

6. Let $A = \{x\}$ and $B = \{x, x\}$. As we already know, these two sets are equal (listing an element of a set more than once is equivalent to listing that element just once). When two sets are equal, they are subsets of each other. That is, if $X = Y$, then $X \subseteq Y$ and $Y \subseteq X$. Conversely, if $X \subseteq Y$ and $Y \subseteq X$, then $X = Y$.

The statement "$X = Y$ if and only if $X \subseteq Y$ and $Y \subseteq X$" is known as the **Axiom of Extensionality**.

Exercise 2.22: For each pair of sets A and B below, determine if $A \subseteq B$, $B \subseteq A$, both, or neither.

1. $A = \{x, y, z\}, B = \{x, z\}$ _____

2. $A = 2\mathbb{Z}, B = \mathbb{N}$ _____

3. $A = \{x, \{x, x\}\}, B = \{\{x\}, x, x\}$ _____

4. $A = 3\mathbb{Z}, B = \{t \in \mathbb{Z} \mid t = 0, 3, 6, 9, …\}$ _____

To the right we see a physical representation of $A \subseteq B$. This figure is called a **Venn diagram**. These types of diagrams are very useful to help visualize relationships among sets. Notice that set A lies completely inside set B. We assume that all the elements of A and B lie in some **universal set** U.

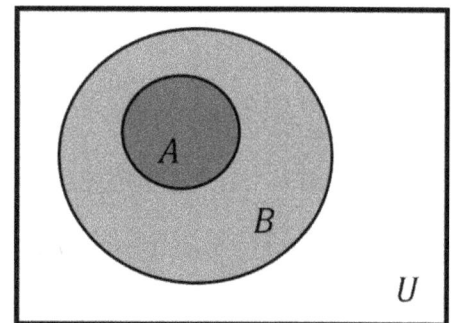

As an example, let U be the set of all species of animals. If we let A be the set of species of cats and we let B be the set of species of mammals, then we have $A \subseteq B \subseteq U$, and we see that the Venn diagram to the right gives a visual representation of this situation. (Note that every cat is a mammal and every mammal is an animal.)

$A \subseteq B$

We say that A is a **proper subset** of B, written $A \subset B$ (or sometimes $A \subsetneq B$), if $A \subseteq B$, but $A \neq B$. For example, $\mathbb{N} \subset \mathbb{Z}$, whereas $\mathbb{N} \not\subset \mathbb{N}$ (although $\mathbb{N} \subseteq \mathbb{N}$).

Note: The definition of proper subset is not too important. It just gives us a convenient way to discuss all the subsets of a specific set except the set itself. For example, it is quite cumbersome to say "Find all subsets of A, but exclude the set A." Instead, we can rephrase this as "Find all proper subsets of A."

The following basic facts about subsets are useful.

Subset Fact 1: Every set is a subset of itself.

Subset Fact 2: The empty set is a subset of every set.

Subset Fact 3: $\subseteq$ is **transitive**. In other words, if $A \subseteq B$ and $B \subseteq C$, then $A \subseteq C$.

Note: To the right we have a Venn diagram illustrating the transitivity of $\subseteq$ (Subset Fact 3).

Since $\subseteq$ is transitive, we can write things like $A \subseteq B \subseteq C \subseteq D$, and without explicitly saying it, we know that $A \subseteq C$, $A \subseteq D$, and $B \subseteq D$.

For example, in part 3 of Example 2.21 above, we saw that

$$\mathbb{N} \subseteq \mathbb{Z} \subseteq \mathbb{Q} \subseteq \mathbb{R} \subseteq \mathbb{C}.$$

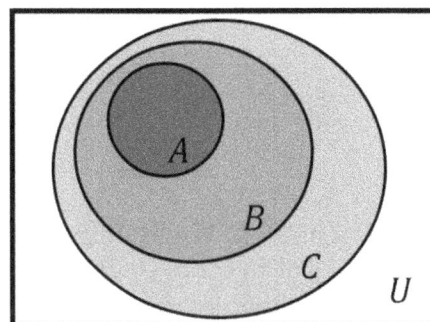

$A \subseteq B \subseteq C$

Since $\subseteq$ is transitive, we automatically know that $\mathbb{N} \subseteq \mathbb{Q}$, $\mathbb{N} \subseteq \mathbb{R}$, $\mathbb{N} \subseteq \mathbb{C}$, $\mathbb{Z} \subseteq \mathbb{R}$, $\mathbb{Z} \subseteq \mathbb{C}$, and $\mathbb{Q} \subseteq \mathbb{C}$.

Example 2.23: Let $C = \{a, b, c\}$, $D = \{a, c\}$, $E = \{b, c\}$, $F = \{b, d\}$, and $G = \emptyset$. Then $D \subseteq C$ and $E \subseteq C$. Also, since *the empty set is a subset of every set*, we have $G \subseteq C$, $G \subseteq D$, $G \subseteq E$, $G \subseteq F$, and $G \subseteq G$. *Every set is a subset of itself*, and so, $C \subseteq C$, $D \subseteq D$, $E \subseteq E$, and $F \subseteq F$.

Exercise 2.24: Draw a Venn Diagram displaying the sets C, D, E, and F from Example 2.23 inside a universal set U.

Power Sets

If A is a set, then the **power set** of A, written $\mathcal{P}(A)$, is the set of all subsets of A. In set-builder notation, we write $\mathcal{P}(A) = \{B \mid B \subseteq A\}$.

Example 2.25: The set $A = \{a, b\}$ has 2 elements and 4 subsets. The subsets of A are $\emptyset$, $\{a\}$, $\{b\}$, and $\{a, b\}$. It follows that $\mathcal{P}(A) = \{\emptyset, \{a\}, \{b\}, \{a, b\}\}$.

The set $B = \{a, b, c\}$ has 3 elements and 8 subsets. The subsets of B are $\emptyset$, $\{a\}$, $\{b\}$, $\{c\}$, $\{a, b\}$, $\{a, c\}$, $\{b, c\}$, and $\{a, b, c\}$. It follows that $\mathcal{P}(B) = \{\emptyset, \{a\}, \{b\}, \{c\}, \{a, b\}, \{a, c\}, \{b, c\}, \{a, b, c\}\}$.

Let's draw a **tree diagram** for the subsets of each of the sets A and B.

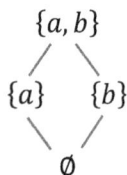

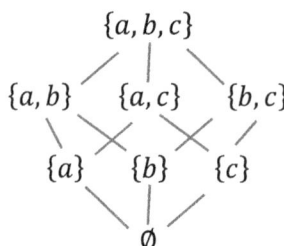

The tree diagram above on the left is for the subsets of the set $A = \{a, b\}$. We start by writing the set $A = \{a, b\}$ at the top. On the next line we write the subsets of cardinality 1 ($\{a\}$ and $\{b\}$). On the line below that we write the subsets of cardinality 0 (just $\emptyset$). We draw a line segment between any two sets when the smaller (lower) set is a subset of the larger (higher) set. So, we see that $\emptyset \subseteq \{a\}$, $\emptyset \subseteq \{b\}$, $\{a\} \subseteq \{a, b\}$, and $\{b\} \subseteq \{a, b\}$. There is actually one more subset relationship, namely $\emptyset \subseteq \{a, b\}$ (and of course each set displayed is a subset of itself). We didn't draw a line segment from $\emptyset$ to $\{a, b\}$ to avoid unnecessary clutter. Instead, we can simply trace the path from $\emptyset$ to $\{a\}$ to $\{a, b\}$ (or from $\emptyset$ to $\{b\}$ to $\{a, b\}$). We are using the **transitivity** of $\subseteq$ here.

The tree diagram above on the right is for the subsets of $B = \{a, b, c\}$. Observe that from top to bottom we write the subsets of B of cardinality 3, then 2, then 1, and then 0. We then draw the appropriate line segments, just as we did for $A = \{a, b\}$.

Exercise 2.26: How many subsets does $\{a, b, c, d\}$ have? _____ Draw a tree diagram for the subsets of $\{a, b, c, d\}$.

Example 2.27:

1. A set with 0 elements must be ∅, and this set has exactly 1 subset (the only subset of the empty set is the empty set itself).

2. A set with 1 element has 2 subsets, namely ∅ and the set itself.

3. In Example 2.25, we saw that a set with 2 elements has 4 subsets, and we saw that a set with 3 elements has 8 subsets.

4. How many subsets does a set of cardinality n have? Do you see the pattern from parts 1, 2, and 3 above? Well, $1 = 2^0$, $2 = 2^1$, $4 = 2^2$, $8 = 2^3$. So, we see that a set with 0 elements has 2^0 subsets, a set with 1 element has 2^1 subsets, a set with 2 elements has 2^2 subsets, and a set with 3 elements has 2^3 subsets.

   It seems that in general, a set with n elements has **2^n** subsets. We can also say that if $|A| = n$, then $|\mathcal{P}(A)| = 2^n$.

Exercise 2.28: Let X be a set such that $|\mathcal{P}(X)| = 128$. What is $|X|$? _____

Basic Set Operations

The **union** of the sets A and B, written $A \cup B$, is the set of elements that are in A or B (or both).

$$A \cup B = \{x \mid x \in A \text{ or } x \in B\}$$

The **intersection** of A and B, written $A \cap B$, is the set of elements that are simultaneously in A and B.

$$A \cap B = \{x \mid x \in A \text{ and } x \in B\}$$

The following Venn diagrams for the union and intersection of two sets can be useful for visualizing these operations.

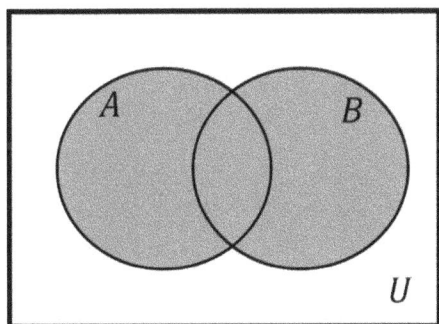

$A \cup B$

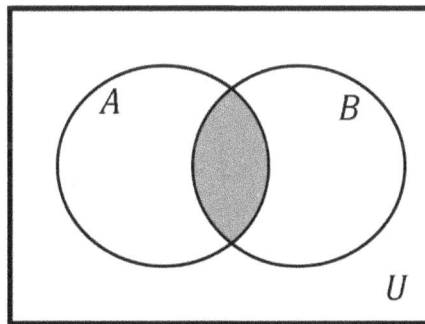

$A \cap B$

The **difference** $A \setminus B$ is the set of elements that are in A and not in B.

$$A \setminus B = \{x \mid x \in A \text{ and } x \notin B\}$$

The **symmetric difference** between A and B, written $A \Delta B$, is the set of elements that are in A or B, but not both.

$$A \Delta B = (A \setminus B) \cup (B \setminus A)$$

41

Let's also look at Venn diagrams for the difference and symmetric difference of two sets.

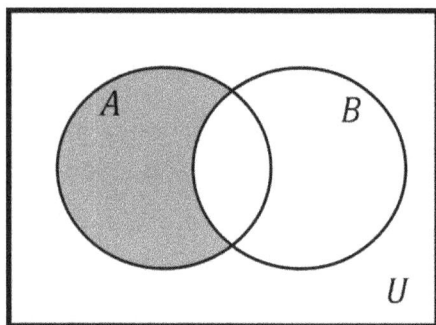

$A \setminus B$

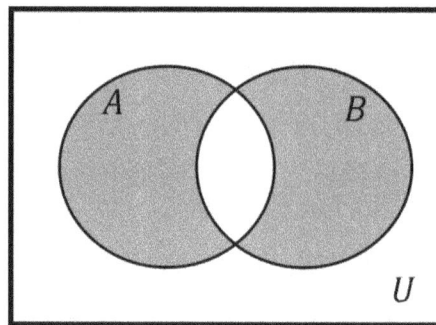

$A \Delta B$

Example 2.29: Let $A = \{0, 1, 2, 3, 4\}$ and $B = \{3, 4, 5, 6\}$. We have

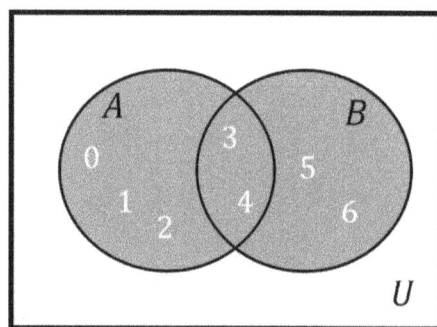

1. $A \cup B = \{0, 1, 2, 3, 4, 5, 6\}$.
2. $A \cap B = \{3, 4\}$.
3. $A \setminus B = \{0, 1, 2\}$.
4. $B \setminus A = \{5, 6\}$.
5. $A \Delta B = \{0, 1, 2\} \cup \{5, 6\} = \{0, 1, 2, 5, 6\}$.

Exercise 2.30: Let $A = \{a, b, \Delta, \delta\}$ and $B = \{b, c, \delta, \gamma\}$. Determine each of the following:

1. $A \cup B$ _____
2. $A \cap B$ _____
3. $A \Delta B$ _____

Example 2.31: Recall that the set of natural numbers is $\mathbb{N} = \{0, 1, 2, 3, \dots\}$ and the set of integers is $\mathbb{Z} = \{\dots, -4, -3, -2, -1, 0, 1, 2, 3, 4, \dots\}$. Observe that in this case, $\mathbb{N} \subseteq \mathbb{Z}$. We have

1. $\mathbb{N} \cup \mathbb{Z} = \mathbb{Z}$.
2. $\mathbb{N} \cap \mathbb{Z} = \mathbb{N}$.
3. $\mathbb{N} \setminus \mathbb{Z} = \emptyset$.
4. $\mathbb{Z} \setminus \mathbb{N} = \{\dots, -4, -3, -2, -1\} = \mathbb{Z}^-$. (Recall that $\mathbb{Z}^-$ is "the set of negative integers.")
5. $\mathbb{N} \Delta \mathbb{Z} = \emptyset \cup \mathbb{Z}^- = \mathbb{Z}^-$.

Note: Whenever A and B are sets and $B \subseteq A$, then

1. $A \cup B = A$.
2. $A \cap B = B$.
3. $B \setminus A = \emptyset$.

Example 2.32: Let $\mathbb{E} = 2\mathbb{N} = \{0, 2, 4, 6, \dots\}$ be the set of even natural numbers and let $\mathbb{O} = 2\mathbb{N} + 1 = \{1, 3, 5, 7, \dots\}$ be the set of odd natural numbers. We have

1. $\mathbb{E} \cup \mathbb{O} = \{0, 1, 2, 3, 4, 5, 6, 7, \dots\} = \mathbb{N}$.
2. $\mathbb{E} \cap \mathbb{O} = \emptyset$.
3. $\mathbb{E} \setminus \mathbb{O} = \mathbb{E}$.
4. $\mathbb{O} \setminus \mathbb{E} = \mathbb{O}$.
5. $\mathbb{E} \triangle \mathbb{O} = \mathbb{E} \cup \mathbb{O} = \mathbb{N}$.

In general, we say that sets A and B are **disjoint** or **mutually exclusive** if $A \cap B = \emptyset$. To the right is a Venn diagram for disjoint sets.

In Example 2.32 above, we saw that the sets $\mathbb{E} = 2\mathbb{N}$ and $\mathbb{O} = 2\mathbb{N} + 1$ are disjoint.

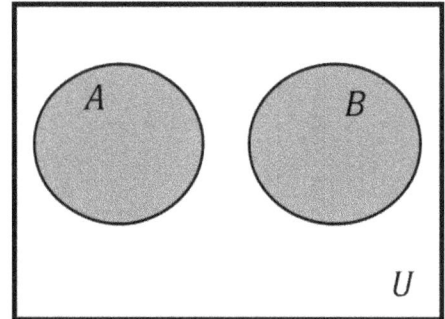

$A \cap B = \emptyset$

Exercise 2.33: Consider the sets $A = \{a + bi \in \mathbb{C} \mid a, b \in \mathbb{Z}\}$ and $B = \{a + bi \in \mathbb{C} \mid a \notin \mathbb{Q}\}$.

1. Is $A \subseteq B$? _____
2. Is $B \subseteq A$? _____
3. Are A and B disjoint? _____

The following basic facts about unions, intersections, and set differences are useful (some of these were mentioned in the Note following Example 2.31).

Set Operation Fact 1: $A \subseteq A \cup B$.

Set Operation Fact 2: $A \cap B \subseteq A$.

Set Operation Fact 3: $B \subseteq A$ if and only if $A \cup B = A$.

Set Operation Fact 4: $B \subseteq A$ if and only if $A \cap B = B$.

Set Operation Fact 5: $B \subseteq A$ if and only if $B \setminus A = \emptyset$.

Exercise 2.34: Show that each of the following statements is false by providing a counterexample.

1. $A \cup B \subseteq A$. $A =$_____ $B =$_____
2. $A \subseteq A \cap B$. $A =$_____ $B =$_____
3. If $B \subseteq A$, then $A \cup B = B$. $A =$_____ $B =$_____
4. If $B \subseteq A$, then $A \setminus B = \emptyset$. $A =$_____ $B =$_____

Unions, intersections, and set differences have many nice algebraic properties such as the following:

1. **Commutativity:** $A \cup B = B \cup A$ and $A \cap B = B \cap A$.
2. **Associativity:** $(A \cup B) \cup C = A \cup (B \cup C)$ and $(A \cap B) \cap C = A \cap (B \cap C)$.
3. **Distributivity:** $A \cap (B \cup C) = (A \cap B) \cup (A \cap C)$ and $A \cup (B \cap C) = (A \cup B) \cap (A \cup C)$.
4. **De Morgan's Laws:** $C \setminus (A \cup B) = (C \setminus A) \cap (C \setminus B)$ and $C \setminus (A \cap B) = (C \setminus A) \cup (C \setminus B)$.
5. **Idempotent Laws:** $A \cup A = A$ and $A \cap A = A$.

Example 2.35: As an example, we can use the following Venn Diagrams to see why the operation of forming unions is associative.

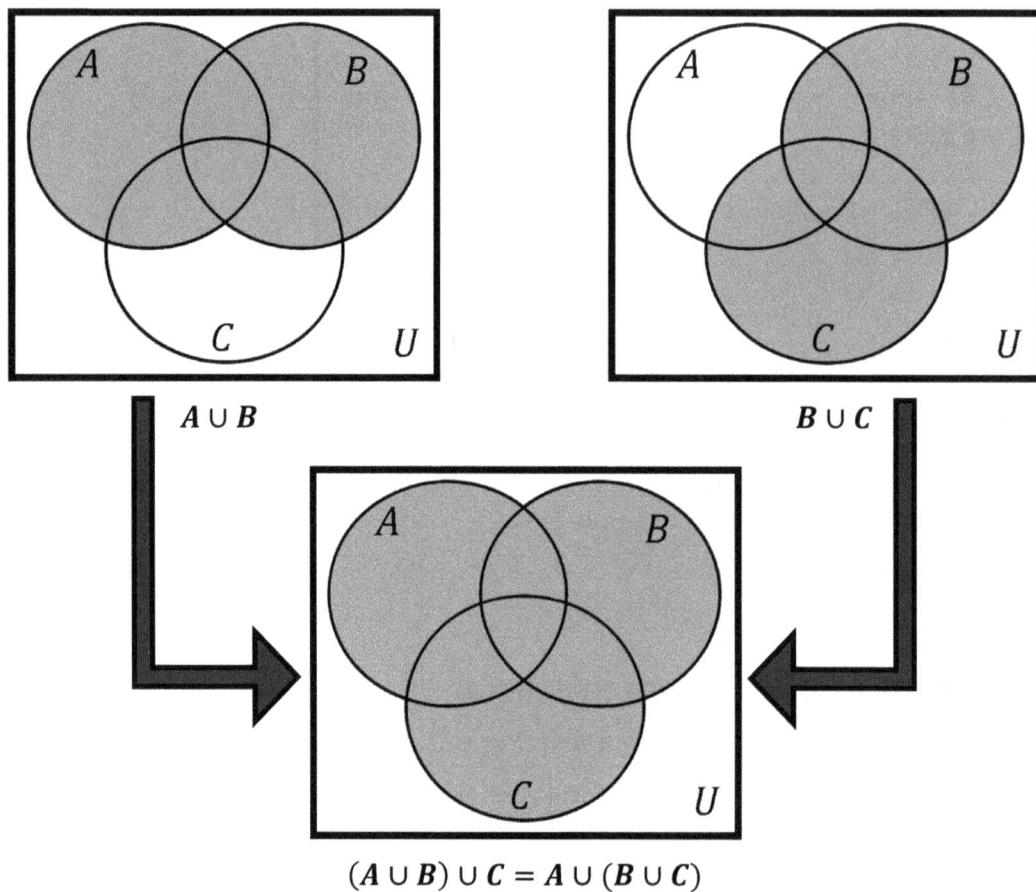

$(A \cup B) \cup C = A \cup (B \cup C)$

The dedicated reader may want to draw Venn Diagrams to help visualize the rest of the properties.

Remember that associativity allows us to drop parentheses. So, we can now simply write $A \cup B \cup C$ when taking the union of the three sets A, B, and C.

Problem Set 2

Full solutions to these problems are available for free download here:
www.SATPrepGet800.com/PMNR2ZX

LEVEL 1

Determine whether each of the following statements is true or false:

1. $b \in \{b\}$

2. $15 \in \{0, 5, 10, 15\}$

3. $-11 \in \{11\}$

4. $0 \in \mathbb{Z}$

5. $-18 \in \mathbb{N}$

6. $\frac{23}{5} \in \mathbb{Q}$

7. $\emptyset \subseteq \{x, y, z, w\}$

8. $\{\Delta\} \subseteq \{\square, \Delta\}$

9. $\{a, b, c, d\} \subset \{a, b, c, d\}$

10. $\{0, 1, \{2, 3\}\} \subseteq \{0, 1, 2, 3\}$

Determine the cardinality of each of the following sets:

11. $\{\text{apple, banana, watermelon}\}$

12. $\{3, 6, 10, 17, 23\}$

13. $\{1, 2, \ldots, 87\}$

14. $\left\{\frac{1}{2}, \frac{1}{3}, \ldots, \frac{1}{10}\right\}$

Provide an example of a set X with the given properties:

15. (i) $X \subset \mathbb{Z}$ (X is a *proper* subset of $\mathbb{Z}$); (ii) X is infinite; (iii) X contains both positive and negative integers; (iv) X contains both even and odd integers.

16. (i) $X \subset \mathbb{R}$ (X is a *proper* subset of $\mathbb{R}$); (ii) X contains both rational and irrational numbers.

17. (i) $X \subset \mathbb{C}$ (X is a *proper* subset of $\mathbb{C}$); (ii) X is infinite; (iii) X contains real numbers; and (iv) X contains complex numbers that are not real.

Let $A = \{x, y, z, w\}$ and $B = \{s, t, y, w\}$. Determine each of the following:

18. $A \cup B$

19. $A \cap B$

20. $A \setminus B$

21. $B \setminus A$

22. $A \, \Delta \, B$

LEVEL 2

Determine whether each of the following statements is true or false:

23. $3 \in \emptyset$

24. $\emptyset \in \emptyset$

25. $\emptyset \in \{\emptyset, \{\emptyset\}\}$

26. $\{\emptyset\} \in \emptyset$

27. $\{\emptyset\} \in \{\emptyset\}$

28. $7 \in \{5k \mid k = 1, 2, 3, 4, 5, 6, 7\}$

29. $13 \in 2\mathbb{N}$

30. $\emptyset \subseteq \emptyset$

31. $\emptyset \subseteq \{\emptyset\}$

32. $\{\emptyset\} \subseteq \emptyset$

33. $\{\emptyset\} \subseteq \{\emptyset\}$

Determine the cardinality of each of the following sets:

34. $\{0, 0, 1, 2, 2, 2, 3, 3\}$

35. $\{\{0, 1\}, \{2, 3, 4\}\}$

36. $\{3, 4, 5, \dots, 2379, 2380\}$

Let $A = \left\{\emptyset, \{\emptyset, \{\emptyset\}\}\right\}$ and $B = \{\emptyset, \{\emptyset\}\}$. Compute each of the following:

37. $A \cup B$

38. $A \cap B$

39. $A \setminus B$

40. $B \setminus A$

41. $A \, \Delta \, B$

Determine if each of the following real numbers is rational or irrational:

42. $1.\overline{3}$

43. $-246.810121416182022242628\dots$

44. $987.65432154321543215432154321\dots$

45. 0

LEVEL 3

Use set-builder notation to describe each of the following sets.

46. $\{1, 3, 5, 7, 9, 11, 13, 15\}$

47. $2\mathbb{N}$

48. $\mathbb{R} \setminus \mathbb{Q}$

Determine the cardinality of each of the following sets:

49. $\{\{\{a, b\}\}\}$

50. $\{\{0, 1\}, 0, \{0\}, \{0, \{0, 1, 2\}\}\}$

51. $\{a, \{a\}, \{a, a\}, \{a, a, a, a\}, \{a, a, \{a\}\}, \{a, \{a\}, \{a\}\}\}$

For each set X, determine $|\mathcal{P}(X)|$

52. $X = \{a, b, c, d, e\}$

53. $X = \{\emptyset, \{\emptyset\}, \{\emptyset, \{\emptyset\}\}\}$

54. $X = \{17, 18, 19, \ldots, 102, 103\}$

Let A, B, and C be sets, let $X = (A \setminus B) \setminus C$, and let $Y = A \setminus (B \setminus C)$.

55. Draw Venn Diagrams for X and Y.

56. Is $X \subseteq Y$?

57. Is $Y \subseteq X$?

58. Is $X = Y$?

LEVEL 4

Determine whether each of the following statements is true or false:

59. $0 \in \{0, \{1\}\}$

60. $\{b\} \in \{a, b\}$

61. $\{1\} \in \{\{1\}, x, 2, y\}$

62. $\emptyset \in \{\{\emptyset\}\}$

63. $\{\{\emptyset\}\} \in \emptyset$

Compute the power set of each of the following sets:

64. $\emptyset$

65. $\{0\}$

66. $\{cat, dog\}$

67. $\{\emptyset, \{\emptyset\}\}$

68. $\{\{\emptyset\}\}$

A **relation** describes a relationship between objects. For example, the relation $=$ describes the relationship "is equal to." Two other relations we have seen are $\in$ (the membership relation) and $\subseteq$ (the subset relation). A relation R is **reflexive** if for all x, we have xRx. A relation R is **symmetric** if for all x, y, we have $xRy \to yRx$. A relation R is **transitive** if for all x, y, z, we have $(xRy \land yRz) \to xRz$. For example, the relation "$=$" is reflexive, symmetric, and transitive because for all x, we have $x = x$, for all x, y, we have $x = y \to y = x$, and for all x, y, z, we have $(x = y \land y = z) \to x = z$.

69. Is $\subseteq$ reflexive?

70. Is $\in$ reflexive?

71. Is $\subseteq$ symmetric?

72. Is $\in$ symmetric?

73. Is $\subseteq$ transitive?

74. Is $\in$ transitive?

LEVEL 5

We say that a set A is **transitive** if every element of A is a subset of A. Determine if each of the following sets is transitive:

75. $\emptyset$

76. $\{\emptyset\}$

77. $\{\{\emptyset\}\}$

78. $\{\emptyset, \{\emptyset\}\}$

79. $\left\{\emptyset, \{\emptyset\}, \{\{\emptyset\}\}\right\}$

80. $\left\{\{\emptyset\}, \{\emptyset, \{\emptyset\}\}\right\}$

81. Assuming that A is transitive, is $\mathcal{P}(A)$ transitive?

Let A and B be sets with $B \subseteq A$. Determine if the following are true or false.

82. $A \cap B = A$

83. $A \setminus B \subseteq A$

84. $B \in \mathcal{P}(A)$

85. $B \subseteq \mathcal{P}(A)$

86. $\mathcal{P}(B) \in \mathcal{P}(A)$

87. $\mathcal{P}(B) \subseteq \mathcal{P}(A)$

CHALLENGE PROBLEMS

88. Let $A = \{a, b, c, d\}$, $B = \{X \mid X \subseteq A \wedge d \notin X\}$, and $C = \{X \mid X \subseteq A \wedge d \in X\}$. Show that there is a natural **one-to-one correspondence** (see definition below) between the elements of B and the elements of C. Then generalize this result to a set with $n + 1$ elements for $n > 0$.

    Definition: A **one-to-one correspondence** between two sets is a pairing so that each element of the first set is matched up with exactly one element of the second set, and vice versa.

89. Explain why $\mathcal{P}(A \cap B) = \mathcal{P}(A) \cap \mathcal{P}(B)$.

90. Determine conditions on sets A and B so that $\mathcal{P}(A \cup B) = \mathcal{P}(A) \cup \mathcal{P}(B)$.

LESSON 3
ABSTRACT ALGEBRA

Binary Operations and Closure

A **binary operation** on a set is a rule that combines two elements of the set to produce another element of the set.

Example 3.1: Let $S = \{0, 1\}$. Multiplication on S is a binary operation, whereas addition on S is **not** a binary operation (here we are thinking of multiplication and addition in the "usual" sense, meaning the way we would think of them in elementary school or middle school).

To see that multiplication is a binary operation on S, observe that $0 \cdot 0 = 0$, $0 \cdot 1 = 0$, $1 \cdot 0 = 0$, and $1 \cdot 1 = 1$. Each of the four computations produces 0 or 1, both of which are in the set S.

We summarize these computations in the following **multiplication table**:

$\cdot$	0	1
0	0	0
1	0	1

The multiplication table works as follows: For $a, b \in S$, we evaluate $a \cdot b$ by taking the entry in the row given by a and the column given by b. For example, the computation "$0 \cdot 1 = 0$" is illustrated below.

$\cdot$	0	1
0	0	0
1	0	1

Multiplication ($\cdot$) is a binary operation on S because the only possible "outputs" are 0 and 1.

$\cdot$	0	1
0	0	0
1	0	1

inputs → (0, 1); outputs ← (0, 0, 0, 1)

To see that addition is **not** a binary operation on S, just note that $1 + 1 = 2$, and $2 \notin S$.

Note that in the multiplication table for addition, the "output" 2 is not equal to either of the "inputs" 0 and 1 as we see below.

$+$	0	1
0	0	1
1	1	2

inputs → (0, 1); this output is not equal to either input ← 2

Note: It may seem odd that I used the expression "multiplication table" for the operation of addition. Nonetheless, "multiplication table" is a general expression that can be used for the table describing any binary operation. It would also be acceptable to call this last table an "addition table" instead.

We will use symbols such as ⋆ or ∘ to represent arbitrary binary operations. If the binary operation is one that we are familiar with, then we will use the appropriate symbol. For example, we will use the symbol "+" when the binary operation is addition, as we did in Example 3.1 above.

Exercise 3.2: Let $S = \{-1, 0, 1\}$. Determine if each of the following is a binary operation on S. Draw the corresponding multiplication tables.

1. Multiplication ($\cdot$) _____

2. Addition ($+$) _____

3. Subtraction ($-$) _____

4. $\star$, where $a \star b = a$ _____

$\cdot$	-1	0	1
-1			
0			
1			

$-$	-1	0	1
-1			
0			
1			

$+$	-1	0	1
-1			
0			
1			

$\star$	-1	0	1
-1			
0			
1			

Some authors refer to a binary operation $\star$ on a set S even when the binary operation is not defined on all pairs of elements $a, b \in S$. We will always refer to these "false operations" as **partial binary operations**.

We say that the set S is **closed** under the partial binary operation $\star$ if whenever $a, b \in S$, we have $a \star b \in S$.

Example 3.3:

1. The set $\{0, 1\}$ is closed under multiplication. This is just another way of saying that multiplication is a binary operation on $\{0, 1\}$. We saw this in Example 3.1 above.

2. The set $\{0, 1\}$ is **not** closed under addition. In other words, addition is a partial binary operation on $\{0, 1\}$ that is **not** a binary operation on $\{0, 1\}$. We saw this in Example 3.1 above.

Exercise 3.4: Let $S = \{-1, 0, 1\}$. Determine if S is closed under each of the following partial binary operations.

1. Multiplication ($\cdot$) _____

2. Addition ($+$) _____

3. Subtraction ($-$) _____

4. $\star$, where $a \star b = a$ _____

Example 3.5: Let $S = \{a, b, c\}$ and define $\star$ using the following multiplication table:

$\star$	a	b	c
a	b	c	c
b	c	a	a
c	a	b	b

1. Is $\star$ a binary operation on S? **Yes it is!** We see this from the table because the only possible "outputs" are a, b, and c.

2. Is S closed under $\star$? **Yes it is!** In fact, saying that S is closed under $\star$ is exactly the same thing as saying that $\star$ is a binary operation on S.

3. Let's compute $b \star c$. Remember that we take the entry in the row given by b and the column given by c, as shown below.

$\star$	a	b	c
a	b	c	c
b	c	a	a
c	a	b	b

The result is $b \star c = a$.

Exercise 3.6: Let $X = \{0, 1, 2, 3\}$ and define $\circ$ using the following multiplication table:

$\circ$	0	1	2	3
0	0	1	2	3
1	1	0	1	3
2	2	2	0	3
3	3	2	1	0

1. Is $\circ$ a binary operation on X? _____

2. Is X closed under $\circ$? _____

3. Compute $2 \circ 3$. _____

4. Compute $3 \circ 2$. _____

Example 3.7:

1. The operation of addition on the set of natural numbers is a binary operation because whenever we add two natural numbers we get another natural number. Here, the set S is $\mathbb{N}$ and the operation $\star$ is $+$. Observe that if $a \in \mathbb{N}$ and $b \in \mathbb{N}$, then $a + b \in \mathbb{N}$. The multiplication table (or addition table in this case) for this example looks as follows:

$+$	0	1	2	3	4	$\cdots$
0	0	1	2	3	4	$\cdots$
1	1	2	3	4	5	$\cdots$
2	2	3	4	5	6	$\cdots$
3	3	4	5	6	7	$\cdots$
4	4	5	6	7	8	$\cdots$
$\vdots$	$\vdots$	$\vdots$	$\vdots$	$\vdots$	$\vdots$	$\ddots$

53

The ellipses in the table indicate that we continue to increase each entry by 1 as we move to the right and as we move down from one entry to the next.

For example, if $a = 1$ and $b = 2$ (both elements of $\mathbb{N}$), then $a + b = 1 + 2 = 3$, and $3 \in \mathbb{N}$. We illustrate this computation in the table below:

+	0	1	2	3	4	$\cdots$
0	0	1	2	3	4	$\cdots$
1	1	2	3	4	5	$\cdots$
2	2	3	4	5	6	$\cdots$
3	3	4	5	6	7	$\cdots$
4	4	5	6	7	8	$\cdots$
$\vdots$	$\vdots$	$\vdots$	$\vdots$	$\vdots$	$\vdots$	$\ddots$

2. The operation of multiplication on the set of positive integers is a binary operation because whenever we multiply two positive integers we get another positive integer. Here, the set S is $\mathbb{Z}^+$ and the operation $\star$ is $\cdot$. Observe that if $a \in \mathbb{Z}^+$ and $b \in \mathbb{Z}^+$, then $a \cdot b \in \mathbb{Z}^+$. The multiplication table for this example looks as follows:

$\cdot$	1	2	3	4	$\cdots$
1	1	2	3	4	$\cdots$
2	2	4	6	8	$\cdots$
3	3	6	9	12	$\cdots$
4	4	8	12	16	$\cdots$
$\vdots$	$\vdots$	$\vdots$	$\vdots$	$\vdots$	$\ddots$

For example, if $a = 3$ and $b = 4$ (both elements of $\mathbb{Z}^+$), then $a \cdot b = 3 \cdot 4 = 12$, and $12 \in \mathbb{Z}^+$. We illustrate this computation in the table below:

$\cdot$	1	2	3	4	$\cdots$
1	1	2	3	4	$\cdots$
2	2	4	6	8	$\cdots$
3	3	6	9	12	$\cdots$
4	4	8	12	16	$\cdots$
$\vdots$	$\vdots$	$\vdots$	$\vdots$	$\vdots$	$\ddots$

3. Let $S = \mathbb{Z}$ and define $\star$ by $a \star b = \min\{a, b\}$, where $\min\{a, b\}$ is the smallest of a or b. Then $\star$ is a binary operation on $\mathbb{Z}$. For example, if $a = -5$ and $b = 3$ (both elements of $\mathbb{Z}$), then $a \star b = -5$, and $-5 \in \mathbb{Z}$. The multiplication table for this example looks as follows:

$\star$	$\cdots$	-1	0	1	2	$\cdots$
$\vdots$	$\ddots$	$\vdots$	$\vdots$	$\vdots$	$\vdots$	$\iddots$
-1	$\cdots$	-1	-1	-1	-1	$\cdots$
0	$\cdots$	-1	0	0	0	$\cdots$
1	$\cdots$	-1	0	1	1	$\cdots$
2	$\cdots$	-1	0	1	2	$\cdots$
$\vdots$	$\iddots$	$\vdots$	$\vdots$	$\vdots$	$\vdots$	$\ddots$

54

4. Subtraction on the set of natural numbers is **not** a binary operation (or equivalently, $\mathbb{N}$ is **not** closed under subtraction). To see this, we just need to provide a single **counterexample**. (Recall that a counterexample is an example that is used to show that a statement is false.) If we let $a = 1$ and $b = 2$ (both elements of $\mathbb{N}$), then we see that $a - b = 1 - 2$ is not an element of $\mathbb{N}$.

Exercise 3.8: For each of the following, determine if $\star$ is a binary operation on S. If it is not, provide a counterexample.

1. $S = \mathbb{Z}; a \star b = a - b$ _____

2. $S = 2\mathbb{N} + 1; a \star b = a + b$ _____

3. $S = \mathbb{N}; a \star b = ab + 5a + 2b$ _____

4. $S = (2\mathbb{Z})^+; a \star b = a^b$ $(a^b = \underbrace{a \cdot a \cdots a}_{b \text{ times}})$ _____

Associativity, Commutativity, and Semigroups

Let $\star$ be a binary operation on a set S. We say that $\star$ is **associative** in S if for all x, y, z in S, we have

$$(x \star y) \star z = x \star (y \star z)$$

A **semigroup** is a pair $(S, \star)$, where S is a set and $\star$ is an associative binary operation on S.

Example 3.9:

1. Let $S = \{a, b\}$ and define $\star$ using the following multiplication table:

$\star$	a	b
a	a	a
b	a	b

Since all four outputs are a or b, we see that $\star$ is a binary operation on S. We would now like to show that $\star$ is associative in S. We need to check that $(x \star y) \star z = x \star (y \star z)$ for all possible combinations of x, y, and z being equal to a and b.

As one sample computation, let's compute $(a \star b) \star b$ and $a \star (b \star b)$. First observe from the table that $a \star b = a$ and $b \star b = b$. So, $(a \star b) \star b = a \star b = a$ and $a \star (b \star b) = a \star b = a$. Since $(a \star b) \star b$ and $a \star (b \star b)$ are both equal to a, they are equal to each other, and we have successfully verified one instance of associativity, namely that $(a \star b) \star b$ and $a \star (b \star b)$ are equal.

It is important to be aware that verifying a single instance of associativity is **not** enough to show that an operation is associative in a set. **All** possible instances must be verified. In this case there are 8 such instances that must be checked. To see that there are eight instances, note that in the equation $(x \star y) \star z = x \star (y \star z)$, there are two choices for x, two choices for y, and two choices for z (each variable can be replaced with a or b). Therefore, the total number of choices for x, y, and z simultaneously is $2 \cdot 2 \cdot 2 = \mathbf{8}$.

Let's now verify all eight instances of associativity.

$$(a \star a) \star a = a \star a = a \qquad\qquad a \star (a \star a) = a \star a = a$$
$$(a \star a) \star b = a \star b = a \qquad\qquad a \star (a \star b) = a \star a = a$$
$$(a \star b) \star a = a \star a = a \qquad\qquad a \star (b \star a) = a \star a = a$$
$$(a \star b) \star b = a \star b = a \qquad\qquad a \star (b \star b) = a \star b = a$$
$$(b \star a) \star a = a \star a = a \qquad\qquad b \star (a \star a) = b \star a = a$$
$$(b \star a) \star b = a \star b = a \qquad\qquad b \star (a \star b) = b \star a = a$$
$$(b \star b) \star a = b \star a = a \qquad\qquad b \star (b \star a) = b \star a = a$$
$$(b \star b) \star b = b \star b = b \qquad\qquad b \star (b \star b) = b \star b = b$$

Observe that for each instance above, the computation on the left results in the same output as the computation on the right. We can now say with certainty that $\star$ is associative in S.

Furthermore, since $\star$ is associative in S, $(S, \star)$ is a semigroup.

It's worth noting that for this example, there is a much less tedious way of verifying that $\star$ is associative in S. Recall the following multiplication table from Example 3.1:

$\cdot$	0	1
0	0	0
1	0	1

This is the multiplication table for ordinary multiplication on the set $\{0, 1\}$. Observe that this table is identical to the one given at the beginning of this example, except that all the names have been changed. In other words, we can transform the original multiplication table into this one by replacing each instance of a with 0 and each instance of b with 1. It follows that we can show that $\star$ is associative in $\{a, b\}$ simply by recognizing that $\cdot$ is associative in $\{0, 1\}$ (ordinary multiplication is associative in $\{0, 1\}$).

2. Let $S = \{a, b, c\}$ and define $\star$ using the following multiplication table (this is the same table from Example 3.5):

$\star$	a	b	c
a	b	c	c
b	c	a	a
c	a	b	b

Notice that $(a \star b) \star c = c \star c = b$ and $a \star (b \star c) = a \star a = b$.

So, $(a \star b) \star c = a \star (b \star c)$. However, once again, this single computation does **not** show that $\star$ is associative in S. In fact, we have the following counterexample: $(a \star c) \star b = c \star b = b$ and $a \star (c \star b) = a \star b = c$. Thus, $(a \star c) \star b \neq a \star (c \star b)$.

So, $\star$ is **not** associative in S, and therefore, $(S, \star)$ is **not** a semigroup.

Exercise 3.10: Each of the following is a multiplication table for a binary operation $\star$ on the set $\{a, b\}$. Determine if $\star$ is associative in S.

1.

$\star$	a	b
a	a	a
b	a	a

2.

$\star$	a	b
a	a	b
b	a	a

3.

$\star$	a	b
a	b	b
b	a	a

4.

$\star$	a	b
a	a	a
b	b	b

Example 3.11:

1. $(\mathbb{N}, +)$, $(\mathbb{Z}, +)$, $(\mathbb{N}, \cdot)$, and $(\mathbb{Z}, \cdot)$ are all semigroups. In other words, the operations of addition and multiplication are both associative in $\mathbb{N}$ and $\mathbb{Z}$.

2. Subtraction is **not** associative in $\mathbb{Z}$. To see this, we just need to provide a single counterexample. If we let $a = 1$, $b = 2$, and $c = 3$, then $(a - b) - c = (1 - 2) - 3 = -1 - 3 = -4$ and $a - (b - c) = 1 - (2 - 3) = 1 - (-1) = 1 + 1 = 2$. Since $-4 \neq 2$, subtraction is not associative in $\mathbb{Z}$. It follows that $(\mathbb{Z}, -)$ is **not** a semigroup.

   Note that $(\mathbb{N}, -)$ is also not a semigroup, but for a different reason. Subtraction is not even a binary operation on $\mathbb{N}$ (see part 4 of Example 3.7).

3. Let $2\mathbb{Z} = \{\dots, -6, -4, -2, 0, 2, 4, 6, \dots\}$ be the set of even integers. When we multiply two even integers together, we get another even integer. It follows that multiplication is a binary operation on $2\mathbb{Z}$. Since multiplication is associative in $\mathbb{Z}$ and $2\mathbb{Z} \subseteq \mathbb{Z}$, it follows that multiplication is associative in $2\mathbb{Z}$ (see the Note below). So, $(2\mathbb{Z}, \cdot)$ is a semigroup.

   Similarly, $(2\mathbb{Z}, +)$ is a semigroup.

4. Let $2\mathbb{Z} + 1 = \{\dots, -5, -3, -1, 1, 3, 5, 7, \dots\}$ be the set of odd integers. Whenever we multiply two odd integers together, we always get another odd integer. It follows that multiplication is a binary operation on $2\mathbb{Z} + 1$. Since multiplication is associative in $\mathbb{Z}$ and $2\mathbb{Z} + 1 \subseteq \mathbb{Z}$, it follows that multiplication is associative in $2\mathbb{Z} + 1$ (once again, see the Note below). So, $(2\mathbb{Z} + 1, \cdot)$ is a semigroup.

   However, $(2\mathbb{Z} + 1, +)$ is **not** a semigroup. Associativity is not the issue here. The problem is that when we add two odd integers, we get an even integer. For example, $1 + 1 = 2$. It follows that $+$ is not a binary operation on $2\mathbb{Z} + 1$.

Note: Associativity is **closed downwards**. By this, we mean that if $\star$ is associative in a set A, and $B \subseteq A$ (B is a **subset** of A), then $\star$ is associative in B.

The reason for this is that the definition of associativity involves only a **universal statement**—a statement that describes a property that is true for all elements without mentioning the existence of any new elements. A universal statement begins with the quantifier $\forall$ ("For all" or "Every") and never includes the quantifier $\exists$ ("There exists" or "There is").

As a simple example, if every object in set A is a fruit, and $B \subseteq A$, then every object in B is a fruit. The universal statement we are referring to might be $\forall x\big(P(x)\big)$, where $P(x)$ is the property "x is a fruit."

In the case of associativity, the universal statement is $\forall x \forall y \forall z\big((x \star y) \star z = x \star (y \star z)\big)$.

Exercise 3.12: Determine if each of the following is a semigroup:

1. $(\mathbb{N}, \star)$, where $a \star b = a$ _____

2. $(\mathbb{Z}^+, \star)$, where $a \star b = a^b$ (see part 4 of Exercise 3.8) _____

3. $(\mathbb{Z}, \star)$, where $a \star b = \min\{a, b\}$ (see part 3 of Example 3.7). **Hint:** You may want to check 6 different "cases" when trying to determine if $(a \star b) \star c = a \star (b \star c)$. For example, one such case would be $a \leq b \leq c$. Another such case would be $a \leq c \leq b$.

Case 1: _____

_____

Case 2: _____

_____

Case 3: _____

_____

Case 4: _____

_____

Case 5: _____

_____

Case 6: _____

_____

Note: Associativity allows us to "add and drop parentheses" as we like. For example, since $+$ is associative in $\mathbb{N}$, the expression "$1 + 3 + 6$" makes sense. Indeed, it doesn't matter if we think about $1 + 3 + 6$ as $(1 + 3) + 6$ or $1 + (3 + 6)$. In each case, we get an answer of 10. More generally, if $\star$ is associative in a set S, then for $a, b, c \in S$, we can write $a \star b \star c$ without worrying about being unclear.

Let $\star$ be a binary operation on a set S. We say that $\star$ is **commutative** (or **Abelian**) in S if for all x, y in S, we have $x \star y = y \star x$.

Example 3.13:

1. Let $S = \{a, b\}$ and define $\star$ using the following multiplication table:

$\star$	a	b
a	a	a
b	a	b

    We saw in part 1 of Example 3.9 that $(S, \star)$ is a semigroup.

    We will now show that $\star$ is commutative in S. This follows from the following computations:

    $$a \star b = a \qquad\qquad b \star a = a$$

    Since $(S, \star)$ is a semigroup and $\star$ is commutative in S, it follows that $(S, \star)$ is a **commutative semigroup**.

    There is a nice visual way to recognize that a binary operation is commutative in a set just by looking at its multiplication table.

$\star$	a	b
a	a	a
b	a	b

    In the table above, we see that entries on opposite sides of the main diagonal are the same. This tells us that $a \star b = b \star a$.

2. $(\mathbb{N},+)$, $(\mathbb{Z},+)$, $(\mathbb{N}, \cdot)$, and $(\mathbb{Z}, \cdot)$ are all commutative semigroups. In other words, the operations of addition and multiplication are both commutative in $\mathbb{N}$ and $\mathbb{Z}$ (as well as being associative).

3. $(2\mathbb{N},+)$, $(2\mathbb{Z},+)$, $(2\mathbb{N},\cdot)$ and $(2\mathbb{Z}, \cdot)$ are commutative semigroups. This follows from part 1 of Example 3.11 together with the fact that commutativity is closed downwards (review the Note following Example 3.11 and then convince yourself that this is true).

Exercise 3.14: Determine if each of the following semigroups is a commutative semigroup:

1. $(\mathbb{N},\star)$, where $a \star b = a$ _____

2. $(\mathbb{Z},\star)$, where $a \star b = \min\{a, b\}$ (see part 3 of Example 3.7).

   Case 1: _____

   Case 2: _____

Identity and Monoids

Let $(S,\star)$ be a semigroup. An element e of S is called an **identity** with respect to the binary operation $\star$ if for all $a \in S$, we have $e \star a = a \star e = a$.

A **monoid** is a semigroup with an identity.

Example 3.15:

1. Let $S = \{a, b\}$ and define $\star$ using the following multiplication table:

$\star$	a	b
a	a	a
b	a	b

   In part 1 of Example 3.13, we saw that $(S,\star)$ is a commutative semigroup.

   We will now show that b is an identity with respect to $\star$. This follows from the following computations:

   $$b \star a = a \qquad\qquad a \star b = a \qquad\qquad b \star b = b$$

   Since $(S,\star)$ is a commutative semigroup and $b \in S$ is an identity with respect to $\star$, it follows that $(S,\star)$ is a commutative monoid.

   There is a nice visual way to recognize an identity just by looking at a multiplication table.

$\star$	a	b
a	a	a
b	a	b

$\star$	a	b
a	a	a
b	a	b

   In the table above on the left, we see that the column under input b is identical to the input column all the way to the left. This shows us that $a \star b = a$ and $b \star b = b$.

   In the table above on the right, we see that the row to the right of input b is identical to the input row all the way up top. This shows us that $b \star a = a$ and (again) $b \star b = b$.

   The two visualizations together show that b is an identity with respect to $\star$.

2. Let $S = \{a, b, c\}$ and define $\star$ using the following multiplication table.

$\star$	a	b	c
a	a	c	c
b	c	c	c
c	c	c	c

$\star$ is a binary operation on S because the only possible outputs are a and c, and both of these outputs are also inputs.

$\star$ is also associative in S because $(x \star y) \star z = c$ and $x \star (y \star z) = c$ unless x, y, and z are all equal to a, in which case we have $(a \star a) \star a = a \star a = a$ and $a \star (a \star a) = a \star a = a$. Therefore, $(S, \star)$ is a semigroup.

$\star$ is also commutative in S, as can be seen by looking at entries on opposite sides of the main diagonal. Therefore, $(S, \star)$ is a commutative semigroup.

However, $(S, \star)$ is **not** a monoid, as there is no identity. To see this, simply observe that there is no row inside the table that matches the input row at the top of the table. In other words, there is no row of the form $a \ b \ c$ (we could have also made the same observation by looking at columns instead of rows). In fact, b does not appear inside the table at all.

Exercise 3.16: Each of the following is the multiplication table for a semigroup. Determine if each of these semigroups is a monoid.

1.

$\star$	a	b	c
a	b	c	b
b	c	b	c
c	b	c	b

2.

$\star$	a	b	c
a	a	a	a
b	a	b	c
c	a	c	b

Example 3.17:

1. $(\mathbb{N}, +)$ and $(\mathbb{Z}, +)$ are commutative monoids with identity 0 (when we add 0 to any integer a, we get a).

2. $(\mathbb{N}, \cdot)$ and $(\mathbb{Z}, \cdot)$ are commutative monoids with identity 1 (when we multiply any integer a by 1, we get a).

3. $(2\mathbb{Z}, +)$ is also a commutative monoid with identity 0. Note that we already saw in part 3 of Example 3.13 that $(2\mathbb{Z}, +)$ is a commutative semigroup.

4. Let A be a nonempty set. Recall from Lesson 2 that $\mathcal{P}(A) = \{B \mid B \subseteq A\}$. Also, if $X, Y \in \mathcal{P}(A)$, then $X \cup Y = \{x \mid x \in X \text{ or } x \in Y\}$. Let's check that $(\mathcal{P}(A), \cup)$ is a commutative monoid.

   If $X, Y \in \mathcal{P}(A)$, then every element of X is in A and every element of Y is in A. It follows that every element of $X \cup Y$ is in A. So, $X \cup Y \in \mathcal{P}(A)$. This shows that $\cup$ is a binary operation on $\mathcal{P}(A)$. By Example 2.35, $\cup$ is associative in $\mathcal{P}(A)$. Finally, $\emptyset$ is an identity for $(\mathcal{P}(A), \cup)$ because if $X \in \mathcal{P}(A)$, then $X \cup \emptyset = X$ and $\emptyset \cup X = X$. It follows that $(\mathcal{P}(A), \cup)$ is a monoid. Since the operation of taking unions is commutative, $(\mathcal{P}(A), \cup)$ is a commutative monoid.

Exercise 3.18: Determine if each of the following is a monoid. If it is not a monoid, is it a semigroup? Is it commutative?

1. $(2\mathbb{N}, +)$ _____

2. $(\mathbb{N}, \star)$, where $a \star b = a$ _____

3. $(\mathbb{Z}, \star)$, where $a \star b = \min\{a, b\}$ (see part 3 of Example 3.7). _____

4. $(2\mathbb{Z}, \cdot)$ _____

5. $(\mathcal{P}(A), \cap)$, where A is a nonempty set. _____

By definition, every monoid has at least one identity. It is natural to ask if a monoid can have more than one identity. We will now discuss why this cannot happen.

The following basic fact about monoids is very useful.

Monoid Fact 1: In any monoid $(M, \star)$, the identity is unique.

Analysis: There is a standard way to show that an object satisfying a certain property (or properties) is unique. You begin by assuming that you have two such objects (not necessarily distinct) and then you argue that they must be the same.

Assume that e and f are both identities.

Since e is an identity, for any $a \in M$, we have $e \star a = a$ and $a \star e = a$. In particular, since $f \in M$, we have $e \star f = f$ and $f \star e = f$.

Since f is an identity, for any $a \in M$, we have $f \star a = a$ and $a \star f = a$. In particular, since $e \in M$, we have $f \star e = e$ and $e \star f = e$.

Now, let's write down all these equations and choose two of them that will show that Monoid Fact 1 is true:

$$\boxed{e \star f = f} \qquad f \star e = f \qquad f \star e = e \qquad \boxed{e \star f = e}$$

The two equations in rectangles show that e and f are equal to the same expression, namely $e \star f$. Therefore, they must be equal to each other. (Note that we could have also used the two equations not in rectangles.)

Exercise 3.19: Let $(M, \star)$ be a monoid and suppose that e and f are (not necessarily distinct identities) of M.

1. Explain why $e \star f = f$.

2. Explain why $e \star f = e$.

Inverses and Groups

Let $(M, \star)$ be a monoid with identity e. An element $a \in M$ is called **invertible** if there is an element $b \in M$ such that $a \star b = b \star a = e$.

A **group** is a monoid in which every element is invertible.

Groups appear so often in mathematics that it's worth taking the time to explicitly spell out the full definition of a group.

A **group** is a pair $(G, \star)$ consisting of a set G together with a binary operation $\star$ satisfying:

(1) **(Associativity)** For all $x, y, z \in G$, $(x \star y) \star z = x \star (y \star z)$.

(2) **(Identity)** There exists an element $e \in G$ such that for all $x \in G$, $e \star x = x \star e = x$.

(3) **(Inverse)** For each $x \in G$, there is $y \in G$ such that $x \star y = y \star x = e$.

Notes: (1) If $y \in G$ is an inverse of $x \in G$, we will usually write $y = x^{-1}$.

(2) Recall that the definition of a binary operation already implies closure. However, many books on groups will mention this property explicitly:

   (Closure) For all $x, y \in G$, $x \star y \in G$.

(3) A group is **commutative** (or **Abelian**) if for all $x, y \in G$, $x \star y = y \star x$.

(4) The properties that define a group are called the **group axioms**. In general, an **axiom** is a statement that is assumed to be true. So, the group axioms are the statements that are **given** to be true in all groups. There are many other statements that are true in groups. However, any additional statements need to be **proved** by using the axioms.

Example 3.20:

1. $(\mathbb{Z}, +)$ is a commutative group with identity 0. The inverse of any integer k is the integer $-k$.

2. $(\mathbb{N}, +)$ is a commutative monoid that is **not** a group. For example, the natural number 1 has no inverse in $\mathbb{N}$. In other words, the equation $x + 1 = 0$ has no solution in $\mathbb{N}$.

Exercise 3.21: Determine if each of the following commutative monoids is a group:

1. $(2\mathbb{Z}, +)$ _____

2. $(2\mathbb{N}, +)$ _____

3. $(\mathbb{Z}, \cdot)$ _____

4. $(\mathcal{P}(A), \cup)$, where A is a nonempty set _____

Example 3.22:

1. Recall that the set of rational numbers is $\mathbb{Q} = \left\{ \frac{a}{b} \mid a, b \in \mathbb{Z} \text{ and } b \neq 0 \right\}$.

   Also, recall that we identify the rational number $\frac{a}{1}$ with the integer a. In this way, $\mathbb{Z} \subseteq \mathbb{Q}$.

We add two rational numbers using the rule $\frac{a}{b} + \frac{c}{d} = \frac{a \cdot d + b \cdot c}{b \cdot d}$.

Note that $0 = \frac{0}{1}$ is an identity for $(\mathbb{Q}, +)$ because $\frac{a}{b} + \frac{0}{1} = \frac{a \cdot 1 + b \cdot 0}{b \cdot 1} = \frac{a}{b}$ and $\frac{0}{1} + \frac{a}{b} = \frac{0 \cdot b + 1 \cdot a}{1 \cdot b} = \frac{a}{b}$.

In fact, $(\mathbb{Q}, +)$ is a commutative group. See parts 1 and 3 of Exercise 3.23 and Problems 55 and 64 below for details.

2. We multiply two rational numbers using the rule $\frac{a}{b} \cdot \frac{c}{d} = \frac{a \cdot c}{b \cdot d}$.

Note that $1 = \frac{1}{1}$ is an identity for $(\mathbb{Q}, \cdot)$ because $\frac{a}{b} \cdot \frac{1}{1} = \frac{a \cdot 1}{b \cdot 1} = \frac{a}{b}$ and $\frac{1}{1} \cdot \frac{a}{b} = \frac{1 \cdot a}{1 \cdot b} = \frac{a}{b}$.

Now, $0 \cdot \frac{a}{b} = \frac{0}{1} \cdot \frac{a}{b} = \frac{0 \cdot a}{1 \cdot b} = \frac{0}{b} = 0$. In particular, when we multiply 0 by any rational number, we can never get 1. So, 0 is a rational number with no multiplicative inverse. It follows that $(\mathbb{Q}, \cdot)$ is **not** a group.

However, 0 is the **only** rational number without a multiplicative inverse. In fact, $(\mathbb{Q}^*, \cdot)$ is a commutative group, where $\mathbb{Q}^*$ is the set of rational numbers with 0 removed ($\mathbb{Q}^* = \mathbb{Q} \setminus \{0\}$). See parts 2 and 4 of Exercise 3.23 and Problems 54 and 63 below for details.

Note: When multiplying two numbers, we will sometimes drop the dot $(\cdot)$ for easier readability. So, we may write $x \cdot y$ as xy. We may also use parentheses instead of the dot. For example, we might write $\frac{a}{b} \cdot \frac{c}{d}$ as $\left(\frac{a}{b}\right)\left(\frac{c}{d}\right)$, whereas we would probably write $\frac{a \cdot c}{b \cdot d}$ as $\frac{ac}{bd}$. We may even use this simplified notation for arbitrary group operations. So, we could write $a \star b$ as ab. However, we will avoid doing this if it would lead to confusion. For example, we will **not** write $a + b$ as ab.

Exercise 3.23: Let $\mathbb{Q}$ be the set of rational numbers and let $\mathbb{Q}^*$ be the set of nonzero rational numbers.

1. Explain why $\mathbb{Q}$ is closed under addition. _____

2. Explain why $\mathbb{Q}^*$ is closed under multiplication. _____

3. If $\frac{a}{b} \in \mathbb{Q}$, then what is the additive inverse of $\frac{a}{b}$? _____

4. If $\frac{a}{b} \in \mathbb{Q}^*$, then what is the multiplicative inverse of $\frac{a}{b}$? _____

Note: Exercise 3.23 verifies two of the five properties of being a commutative group for $(\mathbb{Q}, +)$ and $(\mathbb{Q}^*, \cdot)$, namely **closure** and the **inverse property**. In Example 3.22, we verified the **identity property**. You will be asked to verify **associativity** in Problems 63 and 64 below and **commutativity** in Problems 54 and 55 below.

Example 3.24:

1. Let $C_{12} = \{1, 2, 3, 4, 5, 6, 7, 8, 9, 10, 11, 12\}$. We can define **"clock addition"** by performing addition the way we do it on a 12-hour clock.

   For example, using clock addition, we have $3 + 5 = 8$ and $10 + 3 = 1$ (If it is now $10:00$, then in 3 hours, it will be $1:00$).

$(C_{12}, +)$ is a commutative group with identity 12. It's a bit more natural to name the identity 0 instead of 12, and so, we would normally write $C_{12} = \{0, 1, 2, 3, 4, 5, 6, 7, 8, 9, 10, 11\}$. Note that 1 and 11 are inverses of each other because $1 + 11 = 11 + 1 = 0$. Similarly, we have the following pairs of inverses: $\{2, 10\}, \{3, 9\}, \{4, 8\}, \{5, 7\}$. Also, 0 and 6 are each their own inverse.

There is nothing special about the number 12 here (aside from the fact that many of us own 12-hour clocks). We can consider addition on a "5-hour clock." In this case, we would have $C_5 = \{0, 1, 2, 3, 4\}$ and we can perform computations such as $2 + 3 = 0$ and $3 + 4 = 2$. Once again, $(C_5, +)$ is a commutative group with identity 0. We have the following inverse pairs: $\{1, 4\}$ and $\{2, 3\}$. Once again, 0 is its own inverse.

More generally, for each $n \in \mathbb{Z}^+$, we let $C_n = \{0, 1, 2, \dots, n-1\}$ and we define addition on C_n as if we are adding on an "n-hour clock." Just as in the two special cases above, $(C_n, +)$ is a commutative group with identity 0. If k is between 1 and $n - 1$, inclusive, then the inverse of k is $n - k$ (because $k + (n - k) = 0$). Once again, the inverse of 0 is 0.

2. We can also define "**clock multiplication**" on $C_n = \{0, 1, 2, \dots, n-1\}$ as repeated addition. For example, on C_{12} we have $2 \cdot 5 = 5 + 5 = 10$ and $3 \cdot 7 = (7 + 7) + 7 = 2 + 7 = 9$. We can also perform this last computation by multiplying 3 and 7 as we would "normally" to get $3 \cdot 7 = 21$ and then subtracting 12 to get $21 - 12 = 9$. We can subtract 12 as many times as we need to until we get a number between 0 and 11, inclusive. For example, $5 \cdot 8 = 40$, and subtracting 12 three times gives us a result of $40 - 36 = 4$ (we can also perform this computation more quickly by taking the remainder upon dividing 40 by 12).

Now, $(C_n, \cdot)$ is a commutative monoid with identity 1. However, for each $n \in \mathbb{Z}^+$ with $n > 1$, $(C_n, \cdot)$ is **not** a group. After all, for each $k \in C_n$, we have $0 \cdot k = 0 \neq 1$, and therefore, 0 is **not** invertible.

For some values of n, if we remove 0 from C_n, we will get a group, and sometimes we will not. For example, $(C_3 \setminus \{0\}, \cdot) = (\{1, 2\}, \cdot)$ is a group (where the operation is clock multiplication on a 3-hour clock). 1 and 2 are each their own inverses ($1 \cdot 1 = 1$ and $2 \cdot 2 = 2 + 2 = 1$). As another example, $(C_5 \setminus \{0\}, \cdot) = (\{1, 2, 3, 4\}, \cdot)$ is a group (where the operation is done on a 5-hour clock). 1 and 4 are each their own inverses ($1 \cdot 1 = 1$ and $4 \cdot 4 = 1$), and 2 and 3 are inverses of each other ($2 \cdot 3 = 1$ and $3 \cdot 2 = 1$). However, $(C_4 \setminus \{0\}, \cdot) = (\{1, 2, 3\}, \cdot)$ is **not** a group (where the operation is done on a 4-hour clock). The problem is that $2 \cdot 2 = 0$, and so, the partial binary operation $\cdot$ is not a binary operation on $\{1, 2, 3\}$ (equivalently, $\{1, 2, 3\}$ is not closed under $\cdot$). If we delete the offending element 2 from this set, we see that $(\{1, 3\}, \cdot)$ is a group (where once again, the operation is multiplication modulo 4).

Exercise 3.25: Determine if each of the following is a group. If so, state the inverse of each element. If not, provide a counterexample to show that the inverse property fails.

1. $(C_6, +)$ _____

2. $(C_7 \setminus \{0\}, \cdot)$ _____

3. $(C_{10} \setminus \{0\}, \cdot)$ _____

Example 3.26: Let $G = \{e, a, b\}$, where e, a, and b are distinct, and let $(G, \star)$ be a group with identity e. Let's construct the multiplication table for $(G, \star)$.

Since $e \star x = x \star e = x$ for all x in the group, we can easily fill out the first row and the first column of the table.

$\star$	e	a	b
e	e	a	b
a	a		
b	b		

Next, let's look at the entry corresponding to $a \star a$. We mark this entry with $\boxdot$ below.

$\star$	e	a	b
e	e	a	b
a	a	$\boxdot$	
b	b		

The entry labeled with $\boxdot$ must be either e or b because a is already in that row (see Note 1 below). If it were e, then the final entry in the row would be b, giving two b's in the last column. Therefore, the entry labeled with $\boxdot$ must be b.

$\star$	e	a	b
e	e	a	b
a	a	b	
b	b		

Since the same element cannot be repeated in any row or column (once again, see Note 1 below), the rest of the table is now determined.

$\star$	e	a	b
e	e	a	b
a	a	b	e
b	b	e	a

Notes: (1) Why can't the same element appear twice in any row? Well if x appeared twice in the row corresponding to y, that would mean that there are elements z and w with $z \neq w$ such that $y \star z = x$ and $y \star w = x$. So, $y \star z = y \star w$. Then we have

$$z = e \star z = (y^{-1} \star y) \star z = y^{-1} \star (y \star z) = y^{-1} \star (y \star w) = (y^{-1} \star y) \star w = e \star w = w.$$

This **contradiction** establishes that no element x can appear twice in the same row of a group multiplication table.

We just showed that every group $(G, \star)$ satisfies the **left cancellation law**:

If $a, b, c \in G$ and $ab = ac$, then $b = c$.

A similar argument can be used to show that every group $(G, \star)$ satisfies the **right cancellation law**:

If $a, b, c \in G$ and $ba = ca$, then $b = c$.

By the right cancellation law, the same element in a group cannot appear twice in any column.

(2) The argument given in Note 1 used all the group properties (associativity, identity, and inverse). The dedicated reader should take the time to determine which group property was used for each equality.

What if we remove one of the properties? For example, what about the multiplication table for a monoid? Can an element appear twice in a row or column? I leave this question for the reader to think about (The solution to Problem 10 below will provide an answer).

(3) In Note 1 above, we showed that in the multiplication table for a group, the same element cannot appear as the output more than once in any row or column. We can also show that every element must appear in every row and column. Let's show that the element y must appear in the row corresponding to x. We are looking for an element z such that $x \star z = y$. Well, $z = x^{-1} \star y$ works. Indeed, we have $x \star (x^{-1} \star y) = (x \star x^{-1}) \star y = e \star y = y$.

(4) Using Notes 1 and 3, we see that each element of a group appears exactly once in every row and column of the group's multiplication table.

(5) In the multiplication table we constructed, we can see that e is the identity because the row corresponding to e is the same as the "input row," and the column corresponding to e is the same as the "input column."

$\star$	e	a	b
e	e	a	b
a	a	b	e
b	b	e	a

$\star$	e	a	b
e	e	a	b
a	a	b	e
b	b	e	a

(6) Since $a \star b = e$ and $b \star a = e$, we see that a and b are inverses of each other. Also, since $e \star e = e$, we see that e is its own inverse.

(7) How do we know that $\star$ is associative in G? One way to verify this is by "brute force." This is quite tedious, as there are 27 equalities that need to be verified. For example, $(a \star b) \star b = e \star b = b$ and $a \star (b \star b) = a \star a = b$. Thus, $(a \star b) \star b = a \star (b \star b)$. There are now just 26 more instances of associativity to check.

There is a much less tedious way of verifying that $\star$ is associative in G. We can simply observe that the multiplication table that we constructed is essentially just the multiplication table (or equivalently, the addition table) for clock addition on a 3-hour clock. If we replace e, a, and b with 0, 1, and 2, respectively, and the operation $\star$ with $+$, we get the following table:

$+$	0	1	2
0	0	1	2
1	1	2	0
2	2	0	1

This is just the addition table for $(C_3, +)$. Since we already know that $(C_3, +)$ is associative, it follows that $(G, \star)$ is associative.

(8) We have shown that there is essentially just one group of size 3, namely $(C_3, +)$. Any other group with 3 elements will look exactly like this one, except for possibly the names of the elements. In technical terms, we say that any two groups of order 3 are **isomorphic**.

(9) This group is **commutative**. To see this, we simply observe that entries on opposite sides of the main diagonal are the same.

$\star$	e	a	b
e	e	a	b
a	a	b	e
b	b	e	a

By definition, in a group, each element has at least one inverse. It is natural to ask if a given element in a group can have more than one inverse. As it turns out, this cannot happen! (see Group Fact 1 below.)

The following basic facts about a group $(G, \star)$ are quite useful.

Group Fact 1: Let $a \in G$. Then the inverse of a is unique.

Group Fact 2: To determine if $b \in G$ is the inverse of $a \in G$, we need only show that $a \star b = e$ **or** $b \star a = e$. We do not need to verify both equations.

Exercise 3.27: Let $(G, \star)$ be a group, let $a \in G$, and let b and c be (not necessarily distinct) inverses of a.

1. Explain why $a \star b = e$.

2. Explain why $c \star a = e$.

3. Use parts 1 and 2 above to show that $b = c$.

Warning: The rest of the material in this lesson is on rings and fields, which most students would consider more difficult than the material on semigroups, monoids, and groups. There will be no harm done if you choose to skip ahead to Lesson 4 at this point (after working on some of the problems in Problem Set 3 below) and then come back to review parts of this lesson as needed. Rings and fields will come up briefly in Lesson 5 and then again in Lessons 7 and 8.

Distributivity and Rings

Before giving the general definition of a ring, let's look at an important example.

Example 3.28: Recall that $\mathbb{Z} = \{\ldots, -4, -3, -2, -1, 0, 1, 2, 3, 4, \ldots\}$ is the set of integers. Let's review some of the properties of addition and multiplication on this set.

1. $\mathbb{Z}$ is **closed** under addition. In other words, whenever we add two integers, we get another integer. For example, 2 and 3 are integers, and we have $2 + 3 = 5$, which is also an integer. As another example, -8 and 6 are integers, and so is $-8 + 6 = -2$.

2. Addition is **commutative** in $\mathbb{Z}$. In other words, when we add two integers, it does not matter which one comes first. For example, $2 + 3 = 5$ and $3 + 2 = 5$. So, we see that $2 + 3 = 3 + 2$. As another example, $-8 + 6 = -2$ and $6 + (-8) = -2$. So, we see that $-8 + 6 = 6 + (-8)$.

3. Addition is **associative** in $\mathbb{Z}$. In other words, when we add three integers, it doesn't matter if we begin by adding the first two or the last two integers. For example, $(2 + 3) + 4 = 5 + 4 = 9$ and $2 + (3 + 4) = 2 + 7 = 9$. So, $(2 + 3) + 4 = 2 + (3 + 4)$. As another example, we have $(-8 + 6) + (-5) = -2 + (-5) = -7$ and $-8 + (6 + (-5)) = -8 + 1 = -7$. So, we see that $(-8 + 6) + (-5) = -8 + (6 + (-5))$.

4. $\mathbb{Z}$ has an **identity** for addition, namely 0. Whenever we add 0 to another integer, the result is that same integer. For example, we have $0 + 3 = 3$ and $3 + 0 = 3$. As another example, $0 + (-5) = -5$ and $(-5) + 0 = -5$.

5. Every integer has an additive **inverse**. This is an integer that we add to the original integer to get 0 (the additive identity). For example, the additive inverse of 5 is -5 because we have $5 + (-5) = 0$ and $-5 + 5 = 0$. Notice that the same two equations also show that the inverse of -5 is 5. We can say that 5 and -5 are additive inverses of each other.

We can summarize the five properties above by saying that $(\mathbb{Z}, +)$ is a **commutative group**.

6. $\mathbb{Z}$ is **closed** under multiplication. In other words, whenever we multiply two integers, we get another integer. For example, 2 and 3 are integers, and we have $2 \cdot 3 = 6$, which is also an integer. As another example, -3 and -4 are integers, and so is $(-3)(-4) = 12$.

7. Multiplication is **commutative** in $\mathbb{Z}$. In other words, when we multiply two integers, it does not matter which one comes first. For example, $2 \cdot 3 = 6$ and $3 \cdot 2 = 6$. So, $2 \cdot 3 = 3 \cdot 2$. As another example, $-8 \cdot 6 = -48$ and $6(-8) = -48$. So, we see that $-8 \cdot 6 = 6(-8)$.

8. Multiplication is **associative** in $\mathbb{Z}$. In other words, when we multiply three integers, it doesn't matter if we begin by multiplying the first two or the last two integers. For example, $(2 \cdot 3) \cdot 4 = 6 \cdot 4 = 24$ and $2 \cdot (3 \cdot 4) = 2 \cdot 12 = 24$. So, $(2 \cdot 3) \cdot 4 = 2 \cdot (3 \cdot 4)$. As another example, $(-5 \cdot 2) \cdot (-6) = -10 \cdot (-6) = 60$ and $-5 \cdot (2 \cdot (-6)) = -5 \cdot (-12) = 60$. So, we see that $(-5 \cdot 2) \cdot (-6) = -5 \cdot (2 \cdot (-6))$.

9. $\mathbb{Z}$ has an **identity** for multiplication, namely 1. Whenever we multiply 1 by another integer, the result is that same integer. For example, we have $1 \cdot 3 = 3$ and $3 \cdot 1 = 3$. As another example, $1 \cdot (-5) = -5$ and $(-5) \cdot 1 = -5$.

We can summarize the four properties above by saying that $(\mathbb{Z}, \cdot)$ is a **commutative monoid**.

10. Multiplication is **distributive** over addition in $\mathbb{Z}$. This means that whenever k, m, and n are integers, we have $k \cdot (m + n) = k \cdot m + k \cdot n$. For example, $4 \cdot (2 + 1) = 4 \cdot 3 = 12$ and $4 \cdot 2 + 4 \cdot 1 = 8 + 4 = 12$. So, $4 \cdot (2 + 1) = 4 \cdot 2 + 4 \cdot 1$. As another example, we have $-2 \cdot ((-1) + 3) = -2(2) = -4$ and $-2 \cdot (-1) + (-2) \cdot 3 = 2 - 6 = -4$. Therefore, we see that $-2 \cdot ((-1) + 3) = -2 \cdot (-1) + (-2) \cdot 3$.

Notes: (1) Since the properties listed in 1 through 10 above are satisfied, we say that $(\mathbb{Z}, +, \cdot)$ is a **commutative ring**. We will give the formal definitions of "ring" and "commutative ring" below.

(2) Observe that a ring consists of (i) a set (in this case $\mathbb{Z}$), and (ii) **two** binary operations on the set called **addition** and **multiplication**.

(3) $(\mathbb{Z}, +)$ is a commutative group and $(\mathbb{Z}, \cdot)$ is a commutative monoid. The distributive property is the only property mentioned that requires both addition and multiplication.

(4) We see that $\mathbb{Z}$ is missing one nice property—the inverse property for multiplication. For example, 2 has no multiplicative inverse in $\mathbb{Z}$. There is no integer n such that $2 \cdot n = 1$. So, the linear equation $2n - 1 = 0$ has no solution in $\mathbb{Z}$.

(5) If we replace $\mathbb{Z}$ by the set of natural numbers $\mathbb{N} = \{0, 1, 2, \dots\}$, then all the properties mentioned above are satisfied **except** property 5—the inverse property for addition. For example, 1 has no additive inverse in $\mathbb{N}$. There is no natural number n such that $n + 1 = 0$.

(6) $\mathbb{Z}$ actually satisfies two distributive properties. **Left distributivity** says that whenever $k, m,$ and n are integers, we have $k \cdot (m + n) = k \cdot m + k \cdot n$. **Right distributivity** says that whenever $k, m,$ and n are integers, we have $(m + n) \cdot k = m \cdot k + n \cdot k$. Since multiplication is commutative in $\mathbb{Z}$, left distributivity and right distributivity are equivalent.

For example, if we let $x = 2$, $y = 3$, and $z = 4$, we have

$$2(3 + 4) = 2 \cdot 7 = 14 \text{ and } 2 \cdot 3 + 2 \cdot 4 = 6 + 8 = 14.$$

The picture to the right gives a physical representation of the distributive property for this example. Note that the area of the light grey rectangle is $2 \cdot 3$, the area of the dark grey rectangle is $2 \cdot 4$, and the area of the whole rectangle is $2(3 + 4)$.

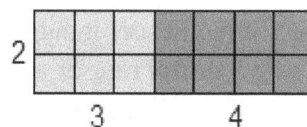

We are now ready to give the more general definition of a ring.

A **ring** is a triple $(R, +, \cdot)$, where R is a set and $+$ and $\cdot$ are binary operations on R satisfying

(1) $(R, +)$ is a commutative group.

(2) $(R, \cdot)$ is a monoid.

(3) Multiplication is **distributive** over addition in R. That is, for all $x, y, z \in R$, we have

$$x \cdot (y + z) = x \cdot y + x \cdot z \qquad \text{and} \qquad (y + z) \cdot x = y \cdot x + z \cdot x.$$

We will always refer to the operation $+$ as addition and the operation $\cdot$ as multiplication. We will also adjust our notation accordingly. For example, we will refer to the identity for $+$ as 0 and the **additive inverse** of an element $x \in R$ as $-x$. Also, we will refer to the identity for $\cdot$ as 1 and the **multiplicative inverse** of an element $x \in R$ (if it exists) as x^{-1} or $\frac{1}{x}$.

Notes: (1) Recall that $(R, +)$ a commutative group means the following:

- **(Closure)** For all $x, y \in R$, $x + y \in R$.

- **(Associativity)** For all $x, y, z \in R$, $(x + y) + z = x + (y + z)$.

- **(Commutativity)** For all $x, y \in R$, $x + y = y + x$.

- **(Identity)** There exists an element $0 \in R$ such that for all $x \in R$, $0 + x = x + 0 = x$.

- **(Inverse)** For each $x \in R$, there is $-x \in R$ such that $x + (-x) = (-x) + x = 0$.

(2) Recall that $(R, \cdot)$ a monoid means the following:

- **(Closure)** For all $x, y \in R$, $x \cdot y \in R$.

- **(Associativity)** For all $x, y, z \in R$, $(x \cdot y) \cdot z = x \cdot (y \cdot z)$.

- **(Identity)** There exists an element $1 \in R$ such that for all $x \in R$, $1 \cdot x = x \cdot 1 = x$.

(3) Although commutativity of multiplication is not required for the definition of a ring, the most well-known example (the ring of integers) satisfies this condition. When multiplication is commutative in R, we call the ring a **commutative ring**. In this case we have the following additional property:

- **(Commutativity)** For all $x, y \in R$, $x \cdot y = y \cdot x$.

(4) Observe that we have two distributive properties in the definition for a ring. The first property is called **left distributivity** and the second is called **right distributivity**.

(5) In a commutative ring, left distributivity implies right distributivity and vice versa. In this case, the distributive property simplifies to

- **(Distributivity)** For all $x, y, z \in R$, $x \cdot (y + z) = x \cdot y + x \cdot z$

(6) Some authors leave out the multiplicative identity property in the definition of a ring and call such a ring a **unital ring** or a **ring with identity**. In this book, we will adopt the convention that a ring has a multiplicative identity. If we do not wish to assume that R has a multiplicative identity, then we will call the structure R "**almost a ring**" or a **rng** (note the missing "i"). If $(R, +, \cdot)$ is "almost a ring," then $(R, \cdot)$ is a semigroup (but not necessarily a monoid).

(7) The properties that define a ring are called the **ring axioms**. In general, an **axiom** is a statement that is assumed to be true. So, the ring axioms are the statements that are **given** to be true in all rings. There are many other statements that are true in rings. However, any additional statements need to be **proved** using the axioms.

Example 3.29:

1. $(\mathbb{Z}, +, \cdot)$ is a commutative ring with additive identity 0 and multiplicative identity 1. The additive inverse of an integer a is the integer $-a$. See Example 3.28 above for more details.

2. $(2\mathbb{Z}, +, \cdot)$ is **not** a ring because $1 \notin 2\mathbb{Z}$. However, it is "almost a ring." In other words, the only property that fails is the multiplicative identity property. In part 1 of Exercise 3.21, you were asked to show that $(2\mathbb{Z}, +)$ is a commutative group and we saw in part 3 of Example 3.11 that $(2\mathbb{Z}, \cdot)$ is a semigroup.

3. $(\mathbb{N}, +, \cdot)$ is **not** a ring because $(\mathbb{N}, +)$ is not a group. The only group property that fails is the additive inverse property. For example, the natural number 1 has no additive inverse. That is, $n + 1 = 0$ has no solution in $\mathbb{N}$. Note that $(\mathbb{N}, \cdot)$ is a commutative monoid and the distributive property holds in $\mathbb{N}$. Therefore, $(\mathbb{N}, +, \cdot)$ misses being a commutative ring by just that one property. $(\mathbb{N}, +, \cdot)$ is an example of a structure called a **semiring**.

4. For each $n \in \mathbb{Z}^+$, $(C_n, +, \cdot)$ is a commutative ring with additive identity 0 and multiplicative identity 1 (here $+$ and $\cdot$ are the operations of clock addition and multiplication on $C_n = \{0, 1, 2, \ldots, n - 1\}$, as defined in Example 3.24).

Note: A semiring $(R, +, \cdot)$ has one additional property called the **zero property**. It says that for all $x \in R$, $0 \cdot x = x \cdot 0 = 0$. We will see below that this property is true in every ring (Ring Fact 1). However, if we take away the additive inverse property, then this property does not follow from the others. So, it must be listed explicitly.

Exercise 3.30: Determine if each of the following is a ring:

1. $(\mathbb{Q}, +, \cdot)$ _____

2. $(\{0\}, +, \cdot)$, where $0 + 0 = 0$ and $0 \cdot 0 = 0$ _____

3. $(\mathbb{Z}^+, +, \cdot)$ _____

4. $(C_{24}, +, \cdot)$ _____

Subtraction and Division: Let $(R, +, \cdot)$ be a ring. If $a, b \in R$, we define $a - b = a + (-b)$ (this is called a **difference**) and if b has a multiplicative inverse in R, then we define $\frac{a}{b} = ab^{-1}$ (this is called a **quotient**).

From experience, one might be led to suspect that multiplying any element of a ring by 0 would result in 0. This turns out to be true. However, this is **not** a ring axiom. Furthermore, it is **not** obvious that this is true. In order to verify that this is true, both the additive inverse property and distributivity are needed. Let's now state this fact formally.

Ring Fact 1: Let $a \in R$. Then $a \cdot 0 = 0$ and $0 \cdot a = 0$.

Analysis: Since 0 is the additive identity of R, we have $0 = 0 + 0$. This allows to write $a \cdot 0$ as $a(0 + 0)$. We can then use distributivity to rewrite $a(0 + 0)$ as $a \cdot 0 + a \cdot 0$. So, we have $a \cdot 0 = a \cdot 0 + a \cdot 0$. Then we have

$$a \cdot 0 = a \cdot 0 + 0 = a \cdot 0 + (a \cdot 0 - a \cdot 0) = (a \cdot 0 + a \cdot 0) - a \cdot 0$$
$$= a(0 + 0) - a \cdot 0 = a \cdot 0 - a \cdot 0 = 0.$$

Exercise 3.31: Let $(R, +, \cdot)$ be a ring with additive identity 0 and let $a \in R$. Explain why $0 \cdot a = 0$.

Example 3.32: Let $R = \{0, 1, x\}$, where 0, 1, and x are distinct, and let $(R, +, \cdot)$ be a ring with additive identity 0 and multiplicative identity 1. Let's construct the addition and multiplication tables for $(R, +, \cdot)$.

By Example 3.26, we already know that the addition table must look as follows (note that we are simply renaming e as 0, a as 1, and b as x):

$+$	0	1	x
0	0	1	x
1	1	x	0
x	x	0	1

Next, let's use Ring Fact 1 to fill in the first row and column of the multiplication table:

$\cdot$	0	1	x
0	0	0	0
1	0		
x	0		

We can now use the fact that 1 is the multiplicative identity to fill in the second row and column of the multiplication table:

$\cdot$	0	1	x
0	0	0	0
1	0	1	x
x	0	x	

Now, since $x = 1 + 1$ (from the addition table), we have

$$x \cdot x = (1 + 1) \cdot x = 1 \cdot x + 1 \cdot x = x + x = 1.$$

It follows that the only possible ring with the given conditions has the following addition and multiplication tables:

+	0	1	x
0	0	1	x
1	1	x	0
x	x	0	1

$\cdot$	0	1	x
0	0	0	0
1	0	1	x
x	0	x	1

It's standard to use the symbol "2" in place of "x," and so, we have the following tables:

+	0	1	2
0	0	1	2
1	1	2	0
2	2	0	1

$\cdot$	0	1	2
0	0	0	0
1	0	1	2
2	0	2	1

Note: To be certain that with the two tables constructed above, $(R, +, \cdot)$ is ring, we need to verify that multiplication is associative in G and that multiplication is distributive over addition in G. This can be done by brute force (doing every computation) or more simply by recognizing that these are the addition and multiplication tables of $(C_3, +, \cdot)$. In other words, these tables represent "clock addition and multiplication" on a 3-hour clock. For the same reason, we see that multiplication is commutative in R, and so, $(R, +, \cdot)$ is a commutative ring.

Fields

A **field** is a triple $(F, +, \cdot)$, where F is a set and $+$ and $\cdot$ are binary operations on F satisfying

(1) $(F, +)$ is a commutative group.

(2) $(F^*, \cdot)$ is a commutative group.

(3) $\cdot$ is **distributive** over $+$ in F. That is, for all $x, y, z \in F$, we have

$$x \cdot (y + z) = x \cdot y + x \cdot z \qquad \text{and} \qquad (y + z) \cdot x = y \cdot x + z \cdot x.$$

(4) $0 \neq 1$.

72

We will refer to the operation $+$ as addition, the operation $\cdot$ as multiplication, the additive identity as 0, the multiplicative identity as 1, the additive inverse of an element $x \in F$ as $-x$, and the multiplicative inverse of an element $x \in F^*$ as x^{-1} or $\frac{1}{x}$. As usual, we will often abbreviate $x \cdot y$ as xy.

Notes: (1) Recall that $(F, +)$ a commutative group means the following:

- **(Closure)** For all $x, y \in F$, $x + y \in F$.

- **(Associativity)** For all $x, y, z \in F$, $(x + y) + z = x + (y + z)$.

- **(Commutativity)** For all $x, y \in F$, $x + y = y + x$.

- **(Identity)** There exists an element $0 \in F$ such that for all $x \in F$, $0 + x = x + 0 = x$.

- **(Inverse)** For each $x \in F$, there is $-x \in F$ such that $x + (-x) = (-x) + x = 0$.

(2) Similarly, $(F^*, \cdot)$ a commutative group means the following:

- **(Closure)** For all $x, y \in F^*$, $xy \in F^*$.

- **(Associativity)** For all $x, y, z \in F^*$, $(xy)z = x(yz)$.

- **(Commutativity)** For all $x, y \in F^*$, $xy = yx$.

- **(Identity)** There exists an element $1 \in F^*$ such that for all $x \in F^*$, $1x = x \cdot 1 = x$.

- **(Inverse)** For each $x \in F^*$, there is $x^{-1} \in F^*$ such that $xx^{-1} = x^{-1}x = 1$.

(3) Recall that F^* is the set of nonzero elements of F. We can write $F^* = \{x \in F \mid x \neq 0\}$ (pronounced "the set of x in F such that x is not equal to 0") or $F^* = F \setminus \{0\}$ (pronounced "F with 0 removed").

(4) The properties that define a field are called the **field axioms**. These are the statements that are **given** to be true in all fields. There are many other statements that are true in fields. However, any additional statements need to be **proved** using the axioms.

(5) If we replace the condition that "$(F^*, \cdot)$ is a commutative group" by "$(F, \cdot)$ is a monoid," then the resulting structure is a ring.

The main difference between a ring and a field is that in a ring, there can be nonzero elements that do not have multiplicative inverses. For example, as we have seen, in the ring $\mathbb{Z}$, 2 has no multiplicative inverse. So, the equation $2x = 1$ has no solution.

(6) Every field is a commutative ring. Although this is not too hard to show (you will be asked to show this in Problem 53 below), it is worth observing that this is not completely obvious. For example, if $(F, +, \cdot)$ is a ring, then since $(F, \cdot)$ is a monoid with identity 1, it follows that $1 \cdot 0 = 0 \cdot 1 = 0$. However, in the definition of a field given above, this property of 0 is not given as an axiom. We **are** given that $(F^*, \cdot)$ is a commutative group, and so, it follows that 1 is an identity for F^*. But $0 \notin F^*$, and so, $1 \cdot 0 = 0 \cdot 1 = 0$ needs to be proved.

Similarly, in the definition of a field given above, 0 is excluded from associativity and commutativity. These need to be checked.

Example 3.33:

1. $(\mathbb{Q}, +, \cdot)$ is a field. In particular, every nonzero element of $\mathbb{Q}$ has a multiplicative inverse. The inverse of the nonzero rational number $\frac{a}{b}$ is the rational number $\frac{b}{a}$. This is easy to verify: $\frac{a}{b} \cdot \frac{b}{a} = \frac{ab}{ba} = \frac{ab}{ab} = \frac{1}{1} = 1$ and $\frac{b}{a} \cdot \frac{a}{b} = \frac{ba}{ab} = \frac{ab}{ab} = \frac{1}{1} = 1$.

2. $(\mathbb{R}, +, \cdot)$ is a field.

3. $(\mathbb{C}, +, \cdot)$ is field, where addition and multiplication of the complex numbers $a + bi$ and $c + di$ are defined as follows:
$$(a + bi) + (c + di) = (a + c) + (b + d)i$$
$$(a + bi)(c + di) = (ac - bd) + (ad + bc)i$$

   The verification of this is not too hard and mostly uses the fact that $(\mathbb{R}, +, \cdot)$ is a field. For example, to verify that addition is commutative in $\mathbb{C}$, we have
$$(a + bi) + (c + di) = (a + c) + (b + d)i = (c + a) + (d + b)i = (c + di) + (a + bi).$$

   We have $a + c = c + a$ because $a, c \in \mathbb{R}$ and addition is commutative in $\mathbb{R}$. For the same reason, we have $b + d = d + b$. You will verify the rest in Problems 67 through 71 below.

4. $(\mathbb{Z}, +, \cdot)$ is a commutative ring that is **not** a field. The only integers with multiplicative inverses are 1 and -1 (they are each their own inverse). In particular, 2 has no multiplicative inverse in $\mathbb{Z}$, or equivalently, the equation $2x = 1$ has no solution in $\mathbb{Z}$.

5. In part 4 of Example 3.29, we saw that for each $n \in \mathbb{Z}^+$, $(C_n, +, \cdot)$ is a commutative ring with additive identity 0 and multiplicative identity 1 (here $+$ and $\cdot$ are the operations of clock addition and multiplication on $C_n = \{0, 1, 2, \ldots, n-1\}$). In part 2 of Example 3.24, we saw that for some values of n, $(C_n \setminus \{0\}, \cdot)$ is a group and for others it is not. For example, if $n = 2$ or $n = 5$, $(C_n \setminus \{0\}, \cdot)$ is a group, and therefore, $(C_2, +, \cdot)$ and $(C_5, +, \cdot)$ are fields. However, for $n = 4$, we saw that $(C_4 \setminus \{0\}, \cdot)$ is **not** a group, and therefore, $(C_4, +, \cdot)$ is not a field. It turns out that $(C_n, +, \cdot)$ is a field if and only if n is a prime number (a prime number is a positive integer that has **exactly** two factors—prime numbers will be discussed in detail in Lesson 4).

6. In Example 3.32, we saw that $(\{0, 1, 2\}, +, \cdot)$ is a commutative ring, where addition and multiplication are defined by the following tables:

+	0	1	2
0	0	1	2
1	1	2	0
2	2	0	1

$\cdot$	0	1	2
0	0	0	0
1	0	1	2
2	0	2	1

   In fact, with these tables, $(\{0, 1, 2\}, +, \cdot)$ is a field. The elements 1 and 2 are each their own multiplicative inverses. Once again, these are the addition and multiplication tables for the field $(C_3, +, \cdot)$.

Exercise 3.34: Determine if each of the following is a field:

1. $(C_{11}, +, \cdot)$ _____

2. $(C_{15}, +, \cdot)$ _____

Full solutions to these problems are available for free download here:

www.SATPrepGet800.com/PMNR2ZX

LEVEL 1

I

$\star$	a	b
a	a	a
b	a	a

II

$\star$	a	b
a	a	b
b	c	a

III

$\star$	a	b
a	a	b
b	b	a

IV

$\star$	a	b
a	a	a
b	b	b

For each of the multiplication tables defined on the set $S = \{a, b\}$ above, determine if each of the following is true or false:

1. $\star$ defines a binary operation on S.

2. $\star$ is commutative in S.

3. a is an identity with respect to $\star$.

4. b is an identity with respect to $\star$.

$\circ$	a	b	c	d
a	a	a	a	b
b	d	d	b	b
c	a	b	c	d
d	c	c	d	d

The multiplication table above is defined on the set $S = \{a, b, c, d\}$.

5. Is S closed under $\circ$?

6. Compute $a \circ b$.

7. Is $\circ$ commutative in S? Why or why not?

8. Is $\circ$ associative in S? Why or why not?

9. Does S have an identity with respect to $\circ$? If so, what is it?

Let $S = \{e, a\}$.

10. Show that there are exactly two monoids on S with identity e.

11. Determine if either of the two monoids on S is a group.

12. Are either of the monoids on S commutative?

LEVEL 2

Define $\star$ on $\mathbb{N}$ by $a \star b = b$.

13. Is $\star$ a binary operation on $\mathbb{N}$?

14. Is $\star$ commutative in $\mathbb{N}$?

15. Is $\star$ associative in $\mathbb{N}$?

16. Does $\mathbb{N}$ have an identity with respect to $\star$? If so, what is it?

17. Is $(\mathbb{N}, \star)$ a semigroup?

18. Is $(\mathbb{N}, \star)$ a monoid?

+	0	1		·	0	1
0	0	1		0	1	0
1	1	0		1	0	1

The addition and multiplication tables above are defined on the set $S = \{0, 1\}$.

19. Does $+$ define a binary operation on S?

20. Does $\cdot$ define a binary operation on S?

21. Does S have an identity with respect to $+$? If so, what is it?

22. Does S have an identity with respect to $\cdot$? If so, what is it?

23. Is $\cdot$ distributive over $+$ in S?

24. Explain why $(S, +, \cdot)$ does **not** define a ring.

+	0	1	2
0	0	1	2
1	1	2	0
2	2	0	1

·	0	1	2
0	0	0	0
1	0	1	2
2	0	2	2

The addition and multiplication tables above are defined on the set $S = \{0, 1, 2\}$.

25. Show that $(S, +, \cdot)$ does **not** define a field.

26. Does $(S, +, \cdot)$ define a ring?

Let $S = \{0, 1\}$ and suppose that $(S, +, \cdot)$ is a ring with additive identity 0 and multiplicative identity 1.

27. Draw the tables for addition and multiplication.

28. Verify that with the tables you drew in Problem 27 that $(S, +, \cdot)$ is a ring.

29. Is $(S, +, \cdot)$ a field?

LEVEL 3

Define $\star$ on $\mathbb{Z}$ by $a \star b = \max\{a, b\}$, where $\max\{a, b\}$ is the largest of a or b.

30. Is $\star$ a binary operation on $\mathbb{Z}$?

31. Is $\star$ commutative in $\mathbb{Z}$?

32. Is $\star$ associative in $\mathbb{Z}$?

33. Does $\mathbb{Z}$ have an identity with respect to $\star$? If so, what is it?

34. Is $(\mathbb{Z}, \star)$ a commutative semigroup?

35. Is $(\mathbb{Z}, \star)$ a commutative monoid?

Let $(G, \star)$ be a group with $G = \{e, a, b, c\}$, where e is the identity. Construct the multiplication table for G under the following conditions:

36. $a^2 = e$ and $b^2 = e$

37. $a^2 = b$ and $b^2 = e$

Let $S = \{-1, 1, -i, i\}$, where i is the complex number such that $i^2 = -1$.

38. Is $+$ a binary operation on S? If so, draw the multiplication table.

39. Is $\cdot$ a binary operation on S? If so, draw the multiplication table.

40. Explain why $\cdot$ is associative in S.

41. Is $(S, \cdot)$ a group? If so, what is the identity, and what is the inverse of each element in S?

If $(R, +, \cdot)$ is a ring, then $a \in R$ is called a **zero divisor** if $a \neq 0$ and there is an element $b \in R$ with $b \neq 0$ such that $ab = 0$. Find all zero divisors in each of the following rings.

42. $(C_5, +, \cdot)$

43. $(C_6, +, \cdot)$

44. $(\mathbb{Z}, +, \cdot)$

LEVEL 4

Let $(G, \star)$ be a group with $a, b \in G$, and let a^{-1} and b^{-1} be the inverses of a and b, respectively. Verify each of the following:

45. $(a \star b)^{-1} = b^{-1} \star a^{-1}$.

46. The inverse of a^{-1} is a.

Let $(F, +, \cdot)$ be a field with $\mathbb{N} \subseteq F$.

47. Explain why $\mathbb{Z} \subseteq F$.

48. Explain why $\left\{ \frac{1}{n} \,\middle|\, n \in \mathbb{Z}^* \right\} \subseteq F$.

49. Explain why $\mathbb{Q} \subseteq F$.

Let $(F, +, \cdot)$ be a field.

50. Show that $\cdot$ is commutative in F.

51. Show that $\cdot$ is associative in F.

52. Show that 1 is a multiplicative identity in F.

53. Explain why $(F, \cdot)$ is a commutative monoid.

Let $\mathbb{Q}$ be the set of rational numbers.

54. Explain why multiplication is commutative in $\mathbb{Q}$.

55. Explain why addition is commutative in $\mathbb{Q}$.

LEVEL 5

Let $(R, +, \cdot)$ be a ring. Explain why each of the following is true:

56. If $a, b \in R$ with $a + b = b$, then $a = 0$.

57. If $a, b \in R$, b^{-1} exists, and $ab = b$, then $a = 1$.

58. If $a, b \in R$, a^{-1} exists, and $ab = 1$, then $b = \frac{1}{a}$.

59. If $a \in R$, then $-a = -1a$.

60. $(-1)(-1) = 1$.

Let $(R, +, \cdot)$ be a ring. Recall that $a \in R$ is called a **zero divisor** if $a \neq 0$ and there is an element $b \in R$ with $b \neq 0$ such that $ab = 0$. A commutative ring with no zero divisors is called an **integral domain**.

61. Give an example of an integral domain that is **not** a field.

62. Explain why every field is an integral domain. Are all rings integral domains?

Let $\mathbb{Q}$ be the set of rational numbers.

63. Explain why multiplication is associative in $\mathbb{Q}$.

64. Explain why addition is associative in $\mathbb{Q}$.

65. Explain why multiplication is distributive over addition in $\mathbb{Q}$.

66. Explain why $(\mathbb{Q}, +, \cdot)$ is a field.

Let $\mathbb{C}$ be the set of complex numbers. To answer the following questions, you may use the fact that $(\mathbb{R}, +, \cdot)$ is a field.

67. Explain why $(\mathbb{C}, +)$ is a commutative group.

68. Explain why multiplication is associative in $\mathbb{C}$.

69. Show that $\mathbb{C}$ has the multiplicative inverse property. What is the inverse of the nonzero complex number $a + bi$?

70. Explain why multiplication is distributive over addition in $\mathbb{C}$.

71. Explain why $(\mathbb{C}, +, \cdot)$ is a field.

Let $S = \{a, b\}$, where $a \neq b$.

72. How many binary operations are there on S?

73. Draw the multiplication table for each binary operation on S.

74. How many semigroups are there of the form $(S, \star)$, up to renaming the elements?

CHALLENGE PROBLEMS

75. Let A be a nonempty set. Prove that $(\mathcal{P}(A), \Delta, \cap)$ is a commutative ring that is not a field (recall that $\mathcal{P}(A)$ is the power set of A and for sets X and Y, $X \Delta Y$ is the symmetric difference between X and Y). Is it an integral domain?

76. Let $(R, +, \cdot)$ be a ring. Recall that $a \in R$ is called a **zero divisor** if $a \neq 0$ and there is an element $b \in R$ with $b \neq 0$ such that $ab = 0$. Also recall that a commutative ring with no zero divisors is called an **integral domain**. Show that every **finite** integral domain is a field.

80

LESSON 4
NUMBER THEORY

Divisibility

An integer a is called **even** if there is another integer b such that $a = 2b$.

Example 4.1:

1. 10 is even because $10 = 2 \cdot 5$. Here $a = 10$ and $b = 5$.

2. -16 is even because $-16 = 2 \cdot (-8)$. Here $a = -16$ and $b = -8$.

3. We can write $1 = 2 \cdot \frac{1}{2}$, but this does **not** show that 1 is even (and as we all know, it is not). In the definition of even, it is very important that b is an integer. The problem here is that $\frac{1}{2}$ is not an integer, and so, it cannot be used as a value for b in the definition of even.

Exercise 4.2: For each integer a, find a number b such that $a = 2b$. Then determine if the equation $a = 2b$ shows that a is even.

1. 6 _____

2. -2 _____

3. 5 _____

4. 0 _____

5. -1 _____

We define the **sum** of integers m and n to be $m + n$. We define the **product** of m and n to be $m \cdot n$.

The following basic facts about even integers are quite useful.

    Even Integer Fact 1: The sum of two even integers is even.

    Even Integer Fact 2: The product of two even integers is even.

Example 4.3:

1. 12 is even because $12 = 2 \cdot 6$ and 16 is even because $16 = 2 \cdot 8$. So, we have
$$12 + 16 = 2 \cdot 6 + 2 \cdot 8 = 2(6 + 8) = 2 \cdot 14 = 28.$$

   For the first equality, we made two substitutions.

   For the second equality, we used the distributivity of multiplication over addition in $\mathbb{Z}$.

   For the third equality, we added the integers 6 and 8 to get the integer 14.

2. Let m and n be even integers. Since m is even, there is an integer j such that $m = 2j$. Since n is even, there is an integer k such that $n = 2k$. So, we have
$$m + n = 2j + 2k = 2(j + k).$$

For the first equality, we made two substitutions.

For the second equality, we used the distributivity of multiplication over addition in $\mathbb{Z}$.

Note that $j + k$ is an integer because the set of integers is closed under addition.

It follows that $m + n$ is even.

Since m and n were arbitrary even integers, we have verified Even Integer Fact 1.

Exercise 4.4: Verify Even Integer Fact 2. **Hint:** Use part 2 of Example 4.3 for guidance.

_____

_____

_____

_____

_____

_____

_____

_____

The property of being even is a special case of the more general notion of divisibility.

An integer a is **divisible** by an integer k, written $k|a$, if there is another integer b such that $a = kb$. We also say that k is a **factor** of a, k is a **divisor** of a, k **divides** a, or a is a **multiple** of k.

Example 4.5:

1. Note that being divisible by 2 is the same as being even. Here $k = 2$.
2. 21 is divisible by 3 because $21 = 3 \cdot 7$. Here $a = 2, b = 7$, and $k = 3$.
3. -48 is divisible by 6 because $-48 = 6 \cdot (-8)$. Here $a = -48, b = -8$, and $k = 6$.

Exercise 4.6: For the integers a and k given, find an integer b that shows that a is divisible by k.

1. $a = 12, k = 3$ _____
2. $a = 12, k = 4$ _____
3. $a = -77, k = 11$ _____
4. $a = 13, k = 13$ _____
5. $a = 24, k = -6$ _____

The following basic facts about divisibility are quite useful.

Divisibility Fact 1: The sum of two integers that are each divisible by k is also divisible by k.

Divisibility Fact 2: The product of two integers that are each divisible by k is also divisible by k.

You will be asked to verify these facts in Problems 76 and 77 below.

Prime Numbers

Before defining a prime number, let's make note of a few basic facts.

Notes: (1) Every integer is divisible by 1. Indeed, if $n \in \mathbb{Z}$, then $n = 1 \cdot n$.

(2) Every integer is divisible by itself. Indeed, if $n \in \mathbb{Z}$, then $n = n \cdot 1$.

(3) It follows from Notes 1 and 2 above that every integer greater than 1 has at least 2 positive integer factors (1 and itself).

A **prime number** is a natural number with **exactly** two positive integer factors.

Notes: (1) An equivalent definition of a prime number is the following: A prime number is an integer greater than 1 that is divisible only by 1 and itself.

(2) An integer greater than 1 that is not prime is called **composite**.

Example 4.7:

1. 0 is **not** prime because every positive integer is a factor of 0. Indeed, if n is a positive integer, then $0 = n \cdot 0$, so that $n|0$.

2. 1 is **not** prime because it has only one positive integer factor: if $1 = kb$ with $k > 0$, then $k = 1$ and $b = 1$.

3. The first ten prime numbers are $2, 3, 5, 7, 11, 13, 17, 19, 23$, and 29.

4. 4 is not prime because $4 = 2 \cdot 2$. In fact, the only even prime number is 2 because by definition, an even integer has 2 as a factor.

5. 9 is the first odd integer greater than 1 that is not prime. Indeed, 3, 5, and 7 are prime, but 9 is not because $9 = 3 \cdot 3$.

6. The first ten composite numbers are $4, 6, 8, 9, 10, 12, 14, 15, 16$, and 18.

Exercise 4.8:

1. What are the first 20 prime numbers? _____

2. What are the first 20 composite numbers? _____

3. Is 91 prime or composite? Why? _____

The following basic facts about prime numbers are very important in mathematics.

　　Prime Number Fact 1: There are infinitely many prime numbers.

　　Prime Number Fact 2: Every integer greater than 1 can be written uniquely as a product of prime numbers, up to the order in which the factors are written.

The second fact is known as **The Fundamental Theorem of Arithmetic**. It is used often in many branches of mathematics.

Exercise 4.9: Use Prime Number Fact 2 to explain why every integer greater than 1 has a prime factor.

When we write an integer n as a product of other integers, we call that product a **factorization** of n. If all the factors in the product are prime, we call the product a **prime factorization** of n.

Example 4.10:

1. $20 = 4 \cdot 5$ is a factorization of 20. This is **not** a prime factorization of 20 because 4 is **not** prime. $20 = 2 \cdot 10$ is another factorization of 20. This example shows that factorizations in general are **not** unique.

2. An example of a prime factorization of 20 is $20 = 2 \cdot 2 \cdot 5$. We can also write this prime factorization as $2 \cdot 5 \cdot 2$ or $5 \cdot 2 \cdot 2$. So, you can see that if we consider different orderings of the factors as different factorizations, then prime factorizations are **not** unique. This is why we say that prime factorizations are unique, **up to the order in which the factors are written**.

3. A prime number is equal to its own prime factorization. In other words, we consider a prime number to be a product of primes with just one factor in the product. For example, the prime factorization of 2 is 2.

Exercise 4.11: Find four factorizations of each of the following positive integers, exactly one of which is a prime factorization.

1. 12 _____ _____ _____ _____

2. 18 _____ _____ _____ _____

3. 100 _____ _____ _____ _____

Since prime factorizations are unique only up to the order in which the factors are written, there can be many ways to write a prime factorization. For example, 10 can be written as $2 \cdot 5$ or $5 \cdot 2$. To make things as simple as possible we will always agree to use the **canonical representation** (or **canonical form**). The word "canonical" is just a fancy name for "natural," and the most natural way to write a prime factorization is in increasing order of primes. So, the canonical representation of 10 is $2 \cdot 5$.

As another example, the canonical representation of 18 is $2 \cdot 3 \cdot 3$. We can tidy this up a bit by rewriting $3 \cdot 3$ as 3^2. So, the canonical representation of 18 is $2 \cdot 3^2$.

If you are new to factoring, you may find it helpful to draw a factor tree.

For example, here is a factor tree for 18:

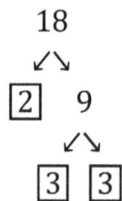

$$18$$
$$\swarrow \searrow$$
$$\boxed{2} \quad 9$$
$$\swarrow \searrow$$
$$\boxed{3} \ \boxed{3}$$

To draw the above tree, we started by writing 18 as the product $2 \cdot 9$. We put a box around 2 because 2 is prime and does not need to be factored any more. We then proceeded to factor 9 as $3 \cdot 3$. We put a box around each 3 because 3 is prime. We now see that we are done, and the prime factorization can be found by multiplying all the boxed numbers together. Remember that we will usually want the canonical representation, and so, we write the final product in increasing order of primes.

By the Fundamental Theorem of Arithmetic (Prime Number Fact 2) above it does not matter how we factor the number—we will always get the same canonical form. For example, here is a different factor tree for 18:

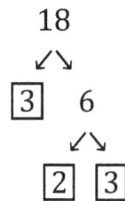

$$18$$
$$\swarrow \searrow$$
$$\boxed{3} \quad 6$$
$$\swarrow \searrow$$
$$\boxed{2} \quad \boxed{3}$$

Now, to show that a positive integer n is composite, we simply need to produce a factor of n that is different from 1 and n itself. This may sound easy, but in practice, as we look at larger and larger values of n it can become very difficult to find factors of n. For example, the largest prime number that we are currently aware of (at the time I am writing this book) is $2^{82,589,933} - 1$. This is an enormous number with 24,862,048 digits. By Prime Number Fact 1, we know that there are prime numbers larger than this, but we have not yet found one.

The following fact about composite numbers provides a couple of tricks to help us (or a computer) determine if a positive integer is prime more quickly.

Composite Number Fact 1: If n is composite, then n has a prime factor $p \leq \sqrt{n}$.

Note: If x and y are real numbers such that $y = x \cdot x$ (or equivalently, $y = x^2$), then we call x a **square root** of y. If x is a positive real number, then we say that x is the **positive square root** of y and we write $x = \sqrt{y}$. For example, $\sqrt{4} = 2$ because $2^2 = 2 \cdot 2 = 4$. As another example, $\sqrt{9} = 3$ because $3^2 = 3 \cdot 3 = 9$.

Example 4.12:

1. Let's determine if 187 is prime or composite. Since $\sqrt{187} < \sqrt{196} = 14$, by Composite Number Fact 1, we need only check to see if 187 is divisible by $2, 3, 5, 7, 11,$ and 13. Checking each of these, we see that $187 = 11 \cdot 17$. So, 187 is composite.

2. Let's determine if 359 is prime or composite. Since $\sqrt{359} < \sqrt{361} = 19$, by Composite Number Fact 1, we need only check to see if 359 is divisible by $2, 3, 5, 7, 11$ 13, and 17. A quick check shows that 359 is **not** divisible by any of these numbers, and so, 359 is prime.

Exercise 4.13: Determine if each of the following positive integers is prime or composite.

1. 119 _____
2. 437 _____
3. 541 _____
4. 1231 _____
5. 1457 _____

We will spend the rest of this section on prime numbers verifying that there are infinitely many prime numbers (Prime Number Fact 1).

If n is a positive integer, then the number $n!$ (pronounced "n **factorial**") is defined by $n! = 1 \cdot 2 \cdots n$.

Example 4.14:

1. $2! = 1 \cdot 2 = 2$.
2. $3! = 1 \cdot 2 \cdot 3 = 6$.
3. $4! = 1 \cdot 2 \cdot 3 \cdot 4 = 24$.

Exercise 4.15: Compute each of the following:

1. $5! = $ ____
2. $6! = $ ____
3. $7! = $ ____

Observe that if $n > 2$, then $n!$ is a number larger than n that is divisible by every positive integer less than or equal to n. For example, $3! = 6$ is divisible by $1, 2$, and 3 (because $3! = 1 \cdot 2 \cdot 3$), and $4! = 24$ is divisible by $1, 2, 3$, and 4 (because $4! = 1 \cdot 2 \cdot 3 \cdot 4$).

So, for $n > 2$, the number $n!$ has many factors. For example, the factors of $3! = 6$ are $1, 2, 3$, and 6. The factors of $4! = 24$ are $1, 2, 3, 4, 6, 8, 12$, and 24.

Exercise 4.16:

1. Find all factors of 5! _____
2. How many factors does 6! have? _____

For $n > 2$, since $n!$ has more than two factors, it cannot be prime. For example, $3!$ has four factors and $4!$ has eight factors. Observe that when we add 1 to $3!$, we get $3! + 1 = 6 + 1 = 7$, an integer that is **not** divisible by 1, 2, or 3. Similarly, when we add 1 to $4!$, we get $4! + 1 = 24 + 1 = 25$, an integer that is **not** divisible by 1, 2, 3, or 4. In general, $n! + 1$ is an integer that is **not** divisible by $1, 2, 3, \ldots, n$.

Exercise 4.17: Let n be any positive integer greater than 2. Use Exercise 4.9 together with the previous paragraph to explain why there is a prime number larger than p.

_____

_____

Note: Exercise 4.17 tells us that there are infinitely many prime numbers.

The Division Algorithm

Before we state the Division Algorithm in its full generality, let's first look at a relatively simple special case.

Recall from the beginning of this lesson that an integer a is **even** if there is another integer b such that $a = 2b$. Additionally, we say that an integer a is **odd** if there is another integer b such that $a = 2b + 1$.

Example 4.18:

1. 6 is even because $6 = 2 \cdot 3$. Here $a = 6$ and $b = 3$.

2. 7 is odd because $7 = 2 \cdot 3 + 1$. Here $a = 7$ and $b = 3$.

3. -6 is even because $-6 = 2(-3)$. Here $a = -6$ and $b = -3$.

4. -7 is odd because $-7 = 2(-4) + 1$. Here $a = -7$ and $b = -4$. Compare this to part 2 above. Based on that example, most students would probably guess that b would turn out to be -3. However, as you can see, that is not the case.

The following fact about even and odd integers is quite useful.

> **The Baby Division Algorithm:** Every integer is even or odd, but not both.

Let's take a moment to restate the Baby Division Algorithm in a way that will more easily generalize to the full Division Algorithm.

> **The Baby Division Algorithm (Alternate Form)**: if a is an integer, then there are unique integers b and r such that $a = 2b + r$, where $r = 0$ or $r = 1$.

Take a moment to convince yourself that the two versions of the Baby Division Algorithm are saying the same thing.

We sometimes say, "When a is divided by 2, b is the **quotient** and r is the **remainder**." Observe that when an integer a is divided by 2, the quotient can be any integer, but the remainder must be 0 or 1.

Example 4.19:

1. When 13 is divided by 2, the quotient is 6 and the remainder is 1. That is, $13 = 2 \cdot 6 + 1$.

2. When 20 is divided by 2, the quotient is 10 and the remainder is 0. That is, $20 = 2 \cdot 10 + 0$, or equivalently, $20 = 2 \cdot 10$. Notice that in this case, 20 is divisible by 2.

3. When – 13 is divided by 2, the quotient is – 7 and the remainder is 1. That is – $13 = 2(-7) + 1$. Compare this to the first example. Based on that example, most students would probably guess that the quotient here would turn out to be – 6. But as you can see, that is not the case.

Exercise 4.20: Determine if a is even or odd. In each case, state the quotient and remainder when a is divided by 2.

1. $a = 46$ _____

2. $a = 97$ _____

3. $a = -38$ _____

4. $a = -51$ _____

The **Division Algorithm** generalizes the notion of an integer a being "even or odd" ($2k$ or $2k + 1$) to a being equal to $mb + r$, where $0 \leq r < m$.

For example, for $m = 3$, the Division Algorithm will tell us that every integer can be written uniquely in one of the three forms $3b$, $3b + 1$, or $3b + 2$. Observe that when an integer a is divided by 3, the quotient can be any integer, but the remainder can be only 0, 1, or 2.

As one more example, for $m = 4$, the Division Algorithm will tell us that every integer can be written uniquely in one of the four forms $4b$, $4b + 1$, $4b + 2$, or $4b + 3$. Observe that when an integer a is divided by 4, the quotient can be any integer, but the remainder can be only 0, 1, 2, or 3.

Example 4.21:

1. When 14 is divided by 3, the quotient is 4 and the remainder is 2. That is, $14 = 3 \cdot 4 + 2$.

2. When 36 is divided by 4, the quotient is 9 and the remainder is 0. That is, $36 = 4 \cdot 9 + 0$, or equivalently, $36 = 4 \cdot 9$. Notice that in this case, 36 is divisible by 4.

3. When 17 is divided by 5, the quotient is 3 and the remainder is 2. That is, $17 = 5 \cdot 3 + 2$.

4. When – 17 is divided by 5, the quotient is – 4 and the remainder is 3. That is – $17 = 5(-4) + 3$.

Exercise 4.22: State the quotient and remainder when a is divided by m.

1. $a = 23, m = 7$ _____

2. $a = 66, m = 11$ _____

3. $a = -49, m = 8$ _____

We now state the full Division Algorithm.

The Division Algorithm: Let a and m be integers with $m > 0$. Then there are unique integers b and r such that $a = mb + r$ with $0 \leq r < m$.

GCD and LCM

Let a and b be two integers. An integer j is a **common divisor** (or **common factor**) of a and b if j is a divisor of both a and b. The **greatest common divisor** (or **greatest common factor**) of a and b, written $\gcd(a, b)$, is the largest common divisor of a and b.

Example 4.23:

1. Let $a = 6$ and $b = 15$. The positive divisors of a are $1, 2, 3$, and 6. The positive divisors of b are $1, 3, 5$, and 15. Therefore, the positive common divisors of a and b are **1 and 3**.

    For each positive divisor there is a corresponding negative divisor. So, a complete list of the divisors of a are $1, 2, 3, 6, -1, -2, -3$, and -6 and a complete list of the divisors of b are $1, 3, 5, 15, -1, -3, -5$, and -15. Therefore, a complete list of the common divisors of a and b are **1, 3, -1, and -3**.

    Wes see that the largest common divisor of a and b is 3. That is, $\gcd(6, 15) = \mathbf{3}$.

    Note that if both a and $-a$ are in a list, we will sometimes use the notation $\pm a$ instead of listing a and $-a$ separately. In this example, we can say that the complete list of common divisors of a and b is $\pm 1, \pm 3$.

2. $\gcd(2, 3) = 1$.

    More generally, if p and q are prime numbers with $p \neq q$, then $\gcd(p, q) = 1$.

3. $\gcd(4, 15) = 1$. Observe that neither 4 nor 15 is prime, and yet their gcd is 1. This is because 4 and 15 have no common factors except for 1 and -1. We say that 4 and 15 are **relatively prime**.

    Note that if p and q are prime numbers with $p \neq q$, then p and q are relatively prime. This example shows that two positive integers can be relatively prime without either integer being prime.

Exercise 4.24: For each of the following, find $\gcd(a, b)$. Are a and b relatively prime?

1. $a = 17, b = 51$ _____
2. $a = 75, b = 90$ _____
3. $a = 19, b = 31$ _____
4. $a = 170, b = 483$ _____

An integer k is a **common multiple** of a and b if k is a multiple of both a and b. The **least common multiple** of a and b, written $\text{lcm}(a, b)$, is the smallest positive common multiple of a and b.

Example 4.25:

1. Let $a = 6$ and $b = 15$. The multiples of a are $\pm 6, \pm 12, \pm 18, \pm 24, \pm 30, \pm 36, \ldots$ and so on. The multiples of 15 are $\pm 15, \pm 30, \pm 45, \pm 60, \ldots$ and so on. Therefore, the common multiples of a and b are $\pm 30, \pm 60, \pm 90, \pm 120, \ldots$ **and so on**.

    It follows that the smallest positive common multiple of a and b is 30. So, $\text{lcm}(6, 15) = \mathbf{30}$.

2. $\text{lcm}(2, 3) = 6$.

   More generally, if p and q are prime numbers with $p \neq q$, then $\text{lcm}(p, q) = pq$.

3. $\text{lcm}(4, 15) = 60$. Observe that neither 4 nor 15 is prime, and yet their lcm is the product of 4 and 15. This is because 4 and 15 are relatively prime (that is, $\gcd(4, 15) = 1$).

   In general, if a and b are relatively prime integers, then $\text{lcm}(a, b) = ab$.

Exercise 4.26: For each of the following, find $\text{lcm}(a, b)$.

5. $a = 17, b = 51$ _____

6. $a = 75, b = 90$ _____

7. $a = 19, b = 31$ _____

8. $a = 170, b = 483$ _____

The following relationship between the gcd and lcm of two positive integers is useful.

    GCD LCM Fact 1: Let $a, b \in \mathbb{Z}^+$. Then $\gcd(a, b) \cdot \text{lcm}(a, b) = ab$.

We can extend the definitions of gcd and lcm to larger sets of numbers as follows:

Let X be a finite set of integers containing at least one nonzero integer. The **greatest common divisor** of the integers in X, written $\gcd(X)$ (or $\gcd(a_1, a_2, \ldots, a_n)$, where $X = \{a_1, a_2, \ldots, a_n\}$) is the largest integer that divides every integer in the set X, and the **least common multiple** of the integers in X, written $\text{lcm}(X)$ (or $\text{lcm}(a_1, a_2, \ldots, a_n)$) is the smallest positive integer that each integer in the set X divides.

For convenience, if X contains only 0, we define $\gcd(X) = 0$ and $\text{lcm}(X) = 0$.

Also, the integers in the set X are said to be **mutually relatively prime** if $\gcd(X) = 1$. The integers in the set X are said to be **pairwise relatively prime** if for each pair $a, b \in X$ with $a \neq b$, $\gcd(a, b) = 1$.

Example 4.27:

1. $\gcd(10, 15, 35) = 5$ and $\text{lcm}(10, 15, 35) = 210$.

2. $\gcd(2, 3, 12) = 1$ and $\text{lcm}(2, 3, 12) = 12$. Notice that here 2, 3, and 12 are mutually relatively prime, but **not** pairwise relatively prime because for example, $\gcd(2, 12) = 2 \neq 1$.

3. $\gcd(10, 21, 143) = 1$ and $\text{lcm}(10, 21, 143) = 30{,}030$. In this case, we have 10, 21, and 143 are pairwise relatively prime.

   We have the following result: if $X = \{a_1, a_2, \ldots, a_n\}$ is a set of pairwise relatively prime integers, then $\gcd(X) = 1$ and $\text{lcm}(X) = a_1 a_2 \cdots a_n$. Also note that a set of pairwise relatively prime integers is mutually relatively prime.

4. For a set X with just one element a, $\gcd(a) = a$ and $\text{lcm}(a) = a$. In particular, $\gcd(0) = 0$ and $\text{lcm}(0) = 0$.

Exercise 4.28: Find the gcd and lcm of each of the following sets of numbers. Determine if the integers in each set are mutually relatively prime, pairwise relatively prime, both, or neither.

1. $\{14, 21, 77\}$ _____   _____   _____

2. $\{2, 3, 5, 7\}$ _____   _____   _____

3. $\{55, 85, 187\}$ _____   _____   _____

4. $\{300, 450, 1470\}$ _____   _____   _____

We will now learn a more systematic method for computing the gcd and lcm of a set of positive integers that uses the prime factorizations of the given integers.

We will slightly modify the canonical representation of a positive integer in a way to ensure that we do not "skip" any primes, and that each prime has a power.

For example, the canonical representation of 50 is $2 \cdot 5^2$. Note that we "skipped over" the prime 3 and there is no exponent written for 2. We can easily give 2 an exponent by rewriting it as 2^1, and since $x^0 = 1$ for any nonzero x (by definition), we can write $1 = 3^0$. Therefore, the prime factorization of 50 can be written as $2^1 \cdot 3^0 \cdot 5^2$.

This convention can be especially useful when comparing two or more positive integers or performing an operation on two or more integers. We will say that $p_0^{a_0} p_1^{a_1} \cdots p_n^{a_n}$ is a **complete prime factorization** if $p_0, p_1, \ldots, p_n$ are the first $n + 1$ primes ($p_0 = 2$, $p_1 = 3$, and so on) and $a_0, a_1, \ldots, a_n \in \mathbb{N}$.

Example 4.29:

1. The prime factorization of 364 in canonical form is $2^2 \cdot 7 \cdot 13$. However, this is **not** a complete prime factorization.

    A complete prime factorization of 364 is $2^2 \cdot 3^0 \cdot 5^0 \cdot 7^1 \cdot 11^0 \cdot 13^1$. This is not the only complete prime factorization of 364. Another one is $2^2 \cdot 3^0 \cdot 5^0 \cdot 7^1 \cdot 11^0 \cdot 13^1 \cdot 17^0$.

    Given a complete prime factorization $p_0^{a_0} p_1^{a_1} \cdots p_n^{a_n}$ of a positive integer, $p_0^{a_0} p_1^{a_1} \cdots p_n^{a_n} p_{n+1}^0$ is another complete prime factorization, and in fact, for any $k \in \mathbb{N}$, $p_0^{a_0} p_1^{a_1} \cdots p_n^{a_n} p_{n+1}^0 p_{n+2}^0 \cdots p_{n+k}^0$ is also a complete prime factorization of that same positive integer. In words, we can include finitely many additional prime factors at the tail end of the original factorization all with exponent 0. Just be careful not to skip any primes!

2. $2^0 \cdot 3^5 \cdot 5^0 \cdot 7^2 \cdot 11^0 \cdot 13^0 \cdot 17^2$ and $2^3 \cdot 3^1 \cdot 5^0 \cdot 7^0 \cdot 11^6$ are complete prime factorizations. In many cases, it is useful to rewrite the second factorization as $2^3 \cdot 3^1 \cdot 5^0 \cdot 7^0 \cdot 11^6 \cdot 13^0 \cdot 17^0$. This is also a complete prime factorization. However, this one has all the same prime factors as the first number given.

Complete prime factorizations give us an easy way to compute greatest common divisors and least common multiples of positive integers.

Suppose that $a = p_0^{a_0} p_1^{a_1} \cdots p_n^{a_n}$ and $b = p_0^{b_0} p_1^{b_1} \cdots p_n^{b_n}$ are complete prime factorizations of a and b. Then we have

$$\gcd(a,b) = p_0^{\min\{a_0,b_0\}} p_1^{\min\{a_1,b_1\}} \cdots p_n^{\min\{a_n,b_n\}} \qquad \mathrm{lcm}(a,b) = p_0^{\max\{a_0,b_0\}} p_1^{\max\{a_1,b_1\}} \cdots p_n^{\max\{a_n,b_n\}}.$$

Example 4.30: Let $a = 2 \cdot 5^2 \cdot 7$ and $b = 3 \cdot 5 \cdot 11^2$. We can rewrite a and b with the following complete prime factorizations: $a = 2^1 \cdot 3^0 \cdot 5^2 \cdot 7^1 \cdot 11^0$ and $b = 2^0 \cdot 3^1 \cdot 5^1 \cdot 7^0 \cdot 11^2$. From these factorizations, it is easy to compute $\gcd(a,b)$ and $\mathrm{lcm}(a,b)$.

$$\gcd(a,b) = 2^0 \cdot 3^0 \cdot 5^1 \cdot 7^0 \cdot 11^0 = 5 \quad \text{and} \quad \mathrm{lcm}(a,b) = 2^1 \cdot 3^1 \cdot 5^2 \cdot 7^1 \cdot 11^2 = 127{,}050.$$

Let's also verify GCD LCM Fact 1 for this example:

$$ab = 2 \cdot 5^2 \cdot 7 \cdot 3 \cdot 5 \cdot 11^2 = 635{,}250 = 5 \cdot 127{,}050 = \gcd(a,b) \cdot \mathrm{lcm}(a,b)$$

Exercise 4.31: For each of the following, find the canonical form of $\gcd(a,b)$ and the canonical form of $\mathrm{lcm}(a,b)$.

1. $a = 2^3 \cdot 3^2, b = 2 \cdot 3 \cdot 5^2$ $\gcd(a,b) =$_____ $\mathrm{lcm}(a,b) =$_____

2. $a = 2^5 \cdot 7^3, b = 3^2 \cdot 5^3$ $\gcd(a,b) =$_____ $\mathrm{lcm}(a,b) =$_____

3. $a = 3872, b = 16{,}731$ $\gcd(a,b) =$_____ $\mathrm{lcm}(a,b) =$_____

4. $a = 12{,}274, b = 122{,}825$ $\gcd(a,b) =$_____ $\mathrm{lcm}(a,b) =$_____

Let a and b be integers. A **linear combination** of a and b is an expression of the form $ma + nb$ with $m, n \in \mathbb{Z}$. We call the integers m and n **weights**.

Example 4.32:

1. Since $5 \cdot 10 - 2 \cdot 15 = 50 - 30 = 20$, we see that 20 is a linear combination of 10 and 15. When we write 20 as $5 \cdot 10 - 2 \cdot 15$, the weights are 5 and -2.

   This is not the only way to write 20 as a linear combination of 10 and 15. For example, we also have $-1 \cdot 10 + 2 \cdot 15 = -10 + 30 = 20$. When we write 20 as $-1 \cdot 10 + 2 \cdot 15$, the weights are -1 and 2.

2. Any number that is a multiple of either 10 or 15 is a linear combination of 10 and 15 because we can allow weights to be 0. For example, 80 is a linear combination of 10 and 15 because $80 = 8 \cdot 10 + 0 \cdot 15$.

   Also, 45 is a linear combination of 10 and 15 because $45 = 0 \cdot 10 + 3 \cdot 15$.

3. It turns out that for any integers a and b, $\gcd(a,b)$ can always be written as a linear combination of a and b. For example, $\gcd(10, 15) = 5$, and we have $5 = -1 \cdot 10 + 1 \cdot 15$.

4. By part 3 above, if a and b are relatively prime, then 1 can be written as a linear combination of a and b. For example, 4 and 15 are relatively prime and we have $1 = 4 \cdot 4 - 1 \cdot 15$.

Parts 3 and 4 of Example 4.32 above lead to the following very useful fact.

Linear Combination Fact 1: Let a and b be integers, at least one of which is not 0. Then $\gcd(a,b)$ is the least positive integer k such that there exist $m, n \in \mathbb{Z}$ with $k = ma + nb$.

Note: Linear Combination Fact 1 says two things:

1. $\gcd(a, b)$ can be written as a linear combination of a and b.

2. A positive integer smaller than $\gcd(a, b)$ **cannot** be written as a linear combination of a and b.

Exercise 4.33: Express k as a linear combination of 12 and 18, or explain why it is not possible.

1. $k = 12$ _____

2. $k = 18$ _____

3. $k = 6$ _____

4. $k = 36$ _____

5. $k = 3$ _____

The Euclidean Algorithm

The Euclidean algorithm is a procedure for computing the gcd of two positive integers. It works very well when it is difficult to find the prime factorization of one or both of the integers. It also provides a method for expressing the gcd as a linear combination of the two given integers.

To avoid confusion, I will explain how to use the Euclidean Algorithm with an example.

Example 4.34: Let's use the Euclidean Algorithm to find $\gcd(70, 126)$.

Step 1: We use the Division Algorithm to find the quotient and remainder when 126 is divided by 70. We have $126 = \mathbf{70} \cdot 1 + \mathbf{56}$. So, the quotient is 1 and the remainder is 56.

Note: Always place the larger of the two positive integers to the left of the equal sign.

Step 2: We now use the Division Algorithm with 70 and 56 (the remainder) to write $70 = \mathbf{56} \cdot 1 + \mathbf{14}$.

Step 3: We now use the Division Algorithm with 56 and 14 to write $56 = 14 \cdot 4 + \mathbf{0}$.

Step 3: Once we get a remainder of 0, the answer is the **previous** remainder. So, $\gcd(70, 126) = \mathbf{14}$.

Exercise 4.35: Use the Euclidean Algorithm to find $\gcd(305, 1040)$.

$$1040 = 305 \cdot \underline{\hspace{1cm}} + \underline{\hspace{1cm}}$$

_____

_____

_____

_____

_____

We can work our way backwards through the Euclidean Algorithm to express $\gcd(a, b)$ as a linear combination of a and b.

Example 4.36: In Example 4.34, we found that $\gcd(70, 126) = 14$ by using the Euclidean Algorithm as follows:

$$126 = 70 \cdot 1 + 56$$
$$70 = 56 \cdot 1 + 14$$
$$56 = 14 \cdot 4 + 0$$

We start with the second to last line (line 2): $70 = 56 \cdot 1 + 14$. We solve this equation for 14 to get $14 = 70 - 56 \cdot 1$.

Working backwards, we next look at line 1: $126 = 70 \cdot 1 + 56$. We solve this equation for 56 and then substitute into the previous equation: $56 = 126 - 70 \cdot 1$. After substituting, we get

$$14 = 70 - 56 \cdot 1 = 70 - (126 - 70 \cdot 1) \cdot 1$$

We then distribute and group all the 70's together and all the 126's together. So, we have

$$14 = 70 - 56 \cdot 1 = 70 - (126 - 70 \cdot 1) \cdot 1 = 70 - 126 \cdot 1 + 70 \cdot 1$$
$$= 70 \cdot 2 - 126 \cdot 1 = 2 \cdot 70 - 1 \cdot 126.$$

So, we see that $\gcd(70, 126) = 14 = \mathbf{2 \cdot 70 - 1 \cdot 126}$, and we have successfully expressed $\gcd(70, 126)$ as a linear combination of 70 and 126.

Exercise 4.37: Use your solution to Exercise 4.35 to express $\gcd(305, 1040)$ as a linear combination of 305 and 1040.

_____

_____

_____

_____

_____

_____

_____

_____

Problem Set 4

LEVEL 1

Write each of the following positive integers as a product of prime factors in canonical form:

1. 16

2. 19

3. 35

4. 105

5. 275

Write each of the following sets using the roster method.

6. $\{n \mid n \text{ is an even prime number}\}$

7. $\{n \mid n \text{ is a prime number less than } 100\}$

8. $\{n \mid n \text{ is one of the first 25 composite numbers}\}$

Find the gcd and lcm of each of the following sets of numbers:

9. $\{6, 9\}$

10. $\{12, 180\}$

11. $\{2, 3, 5\}$

For each of the integers a and k given, find a number b that shows that a is divisible by k.

12. $a = 8, k = 2$

13. $a = 30, k = 3$

14. $a = 17, k = 17$

15. $a = 1006, k = 1$

LEVEL 2

Write each of the following positive integers as a product of prime factors in canonical form:

16. 693

17. 67,500

18. 384,659

19. 9,699,690

Determine if each of the following positive integers is prime:

20. 20

21. 53

22. 71

23. 81

24. 85

25. 93

For $n \in \mathbb{Z}^+$, let $M_n = n! + 1$. Determine if M_n is prime for each of the following values of n.

26. $n = 3$

27. $n = 4$

28. $n = 5$

29. $n = 6$

30. $n = 7$

Find the gcd and lcm of each of the following sets of numbers:

31. $\{14, 21, 77\}$

32. $\{720, 2448, 5400\}$

33. $\{2^{17} \cdot 5^4 \cdot 11^9 \cdot 23, 2^5 \cdot 3^2 \cdot 7^4 \cdot 11^3 \cdot 13\}$

Determine if each of the following numbers is prime:

34. 101

35. 399

36. 1829

37. 1933

38. 8051

39. 13,873

40. 65,623

Use the Division Algorithm to find the quotient and remainder when

41. 28 is divided by 3.

42. 522 is divided by 6.

43. 723 is divided by 17.

44. 2365 is divided by 71.

Express k as a linear combination of 70 and 100, or explain why it is not possible.

45. $k = 100$

46. $k = 70$

47. $k = 10$

48. $k = 5$

49. $k = 1$

For each of the following, use the Euclidean Algorithm to find $\gcd(a, b)$.

50. $a = 15, b = 40$

51. $a = 36, b = 120$

52. $a = 55, b = 300$

53. $a = 825, b = 2205$

LEVEL 4

For each of the following, use your computations from Problems 50 through 53 to express $\gcd(a, b)$ as a linear combination of a and b.

54. $a = 15, b = 40$

55. $a = 36, b = 120$

56. $a = 55, b = 300$

57. $a = 825, b = 2205$

A **prime pair** is a pair of prime numbers of the form $p, p + 2$. For example, $3, 5$ is a prime pair.

58. Is $5, 7$ a prime pair?

59. Is $17, 19$ a prime pair?

60. Is $1001, 1003$ a prime pair?

61. Find a prime pair $p, p + 2$ such that $p > 2000$.

Determine how many factors each of the following positive integers has.

62. 50

63. 1000

64. $7!$

65. $12!$

LEVEL 5

Verify each of the following:

66. If $a, b, c \in \mathbb{Z}$ with $a|b$ and $b|c$, then $a|c$.

67. $\gcd(a, b) \cdot \text{lcm}(a, b) = ab$.

68. If $a, b, c, d, e \in \mathbb{Z}$ with $a|b$ and $a|c$, then $a|(db + ec)$.

If $a, b \in \mathbb{Z}^+$ and $\gcd(a, b) = 1$, find each of the following:

69. $\gcd(a, a + 1)$

70. $\gcd(a, a + 2)$

71. $\gcd(3a + 2, 5a + 3)$

72. $\gcd(a + b, a - b)$

Verify each of the following:

73. If n is composite, then n has a prime factor $p \leq \sqrt{n}$ (this is Composite Number Fact 1).

74. The sum of two odd integers is an even integer.

75. The product of an even integer and any other integer is an even integer.

76. The sum of two integers that are each divisible by k is also divisible by k.

77. The product of an integer divisible by k and any other integer is divisible by k.

78. The product of two odd integers is an odd integer.

CHALLENGE PROBLEMS

79. Show that if $a|c$ and $b|c$, then $\text{lcm}(a, b) \mid c$.

80. A **prime triple** is a sequence of three prime numbers of the form p, $p + 2$, and $p + 4$. For example, $3, 5, 7$ is a prime triple. Show that there are no other prime triples.

81. Find all subgroups of $(\mathbb{Z}, +)$ and all submonoids of $(\mathbb{Z}, +)$.

LESSON 5
REAL ANALYSIS

Ordered Sets

Recall that a **relation** on a set describes a relationship among the set's objects. We will be interested only in **binary relations**. The idea is that given **two** objects from a set, either these two objects satisfy the given relationship or they do not.

Example 5.1:

1. "$=$" (equals) is a binary relation on the set of natural numbers. The statement $0 = 0$ (pronounced "0 is equal to 0") is true, whereas the statement $0 = 1$ is false. Notice that for natural numbers a and b, the statement $a = b$ is true if and only if a and b are the same natural number. Instead of saying "$a = b$ is true," we will simply write $a = b$. Instead of saying "$a = b$ is false," we will write $a \neq b$.

   More generally, if A is any set, then "$=$" is a binary relation on A. Once again, $a = b$ if and only if a and b are the same element of A.

2. Let $A = \{a, b, c\}$. Recall that the power set of A is
$$\mathcal{P}(A) = \{\emptyset, \{a\}, \{b\}, \{c\}, \{a, b\}, \{a, c\}, \{b, c\}, \{a, b, c\}\}.$$
   "$\subseteq$" (subset) is a binary relation on $\mathcal{P}(A)$. We have for example $\{a\} \subseteq \{a, c\}$ (pronounced "$\{a\}$ is a subset of $\{a, c\}$"), whereas $\{a\} \nsubseteq \{b, c\}$.

   More generally, if X is any set, then "$\subseteq$" is a binary relation on $\mathcal{P}(X)$.

3. "$\leq$" (less than or equal to) is a binary relation on the set of natural numbers. We have for example $2 \leq 7$ (pronounced "2 is less than or equal to 7") and $2 \leq 2$, whereas $2 \nleq 1$.

   "$\leq$" is also a binary relation on many other sets such as the integers, the rational numbers, and the real numbers.

Exercise 5.2: Consider the binary relations "$<$" (less than) and "$>$" (greater than) on the set of integers. Determine if each of the following statements is true or false.

1. $0 < -1$ ____
2. $-1 > -3$ ____
3. $-1 \nless -5$ ____
4. $5 \ngtr -5$ ____

We will usually use the letter R to represent an arbitrary binary relation. We can read the statement aRb as "a is related to b." We are going to be interested in binary relations that have certain properties. We will describe some of the most important properties now.

We say that a binary relation R on a set A is

- **reflexive** if for all $a \in A$, aRa.

- **symmetric** if for all $a, b \in A$, aRb implies bRa.

- **transitive** if for all $a, b, c \in A$, aRb and bRc imply aRc.

- **antireflexive** if for all $a \in A$, $a\cancel{R}a$.

- **antisymmetric** if for all $a, b \in A$, aRb and bRa imply $a = b$.

- **trichotomous** if for all $a, b \in A$, exactly one of aRb, bRa, or $a = b$ holds.

Example 5.3:

1. Let A be a nonempty set and consider the binary relation "$=$" on A. This relation is reflexive ($a = a$), symmetric (if $a = b$, then $b = a$), transitive (if $a = b$ and $b = c$, then $a = c$), and antisymmetric (trivially). This relation is **not** antireflexive because $a \neq a$ is false for any $a \in A$. This relation is **not** trichotomous because if $a \neq b$, then $b \neq a$.

2. Let A be a set with at least two elements and consider the binary relation "$\subseteq$" on $\mathcal{P}(A)$. This relation is reflexive (every set is a subset of itself), transitive (if $X \subseteq Y$ and $Y \subseteq Z$, then $X \subseteq Z$), and antisymmetric (if $X \subseteq Y$ and $Y \subseteq X$, then $X = Y$). This relation is **not** symmetric (for example, if a and b are distinct, then $\{a\} \not\subseteq \{b\}$ and $\{b\} \not\subseteq \{a\}$), it is **not** antireflexive ($X \not\subseteq X$ is false for any set X), and it is **not** trichotomous (for example, if a and b are distinct, then $\{a\} \not\subseteq \{b\}$, $\{b\} \not\subseteq \{a\}$, and $\{a\} \neq \{b\}$).

3. Consider the binary relation "$\leq$" on $\mathbb{N}$. This relation is reflexive ($a \leq a$), transitive (if $a \leq b$ and $b \leq c$, then $a \leq c$), and antisymmetric (if $a \leq b$ and $b \leq a$, then $a = b$). This relation is **not** symmetric (for example, $0 \leq 1$, but $1 \not\leq 0$), it is **not** antireflexive (for example $0 \not\leq 0$ is false), and it is **not** trichotomous ($0 \leq 0$ and $0 = 0$ both hold).

Exercise 5.4: Consider the relation $>$ on $\mathbb{Z}$.

1. Is $>$ reflexive on $\mathbb{Z}$? _____

2. Is $>$ symmetric on $\mathbb{Z}$? _____

3. Is $>$ transitive on $\mathbb{Z}$? _____

4. Is $>$ antireflexive on $\mathbb{Z}$? _____

5. Is $>$ antisymmetric on $\mathbb{Z}$? _____

6. Is $>$ trichotomous on $\mathbb{Z}$? _____

A binary relation $<$ on a set A is a **strict linear ordering** on A if $<$ is transitive and trichotomous on A. In this case, we will call $(A, <)$ an **ordered set**.

Example 5.5:

1. The binary relation $<$ on $\mathbb{N}$ is a strict linear ordering. Therefore, $(\mathbb{N}, <)$ is an ordered set.

2. The binary relation $<$ on $\mathbb{Z}$ is a strict linear ordering. Therefore, $(\mathbb{Z}, <)$ is an ordered set.

3. The binary relation $<$ on $\mathbb{Q}$ is a strict linear ordering. Here we have $\frac{a}{b} < \frac{c}{d}$ if and only if $ad < bc$. Therefore, $(\mathbb{Q}, <)$ is an ordered set.

4. The binary relation $=$ is **not** a strict linear ordering on any nonempty set A because it is not trichotomous on A (see part 1 of Example 5.3). Therefore, $(A, =)$ is **not** an ordered set.

5. Let X be the set of planets in our solar system. Define a binary relation $\lhd$ on X by $x \lhd y$ if x is closer to the sun than y. Then $\lhd$ is a strict linear ordering on X. Therefore, $(X, \lhd)$ is an ordered set.

Exercise 5.6: Determine if each of the following is an ordered set.

1. $(\mathbb{N}, \leq)$ _____

2. $(\mathbb{Z}, \leq)$ _____

3. $(\mathbb{N}, >)$ _____

4. $(\mathbb{Z}, >)$ _____

5. $(\mathcal{P}(A), \subset)$, where A is a set with at least two elements and $\subset$ is the proper subset relation.

_____

If $(A, <)$ is an ordered set (so that $<$ is a strict linear ordering on A), then we define the binary relation $\leq$ on A by $a \leq b$ if and only if $a < b$ or $a = b$. In this case, $\leq$ is called a **linear ordering** on A.

Exercise 5.7: Let $(A, <)$ be an ordered set (so that $<$ is a strict linear ordering on A). Verify each of the following statements about the corresponding linear ordering $\leq$.

1. $\leq$ is **not** trichotomous on A. _____

2. $\leq$ is reflexive on A. _____

3. $\leq$ is antisymmetric on A. _____

4. $\leq$ is transitive on A. _____

5. $\leq$ satisfies the following **comparability condition** on A: if $a, b \in A$, then $a \leq b$ or $b \leq a$.

_____

If $(A, <)$ is an ordered set, we may write $b > a$ in place of $a < b$. We may also write $b \geq a$ in place of $a \leq b$.

Ordered Rings and Fields

An **ordered ring** is a quadruple $(R, +, \cdot, <)$, where $(R, +, \cdot)$ is a ring, and $(R, <)$ is an ordered set such that:

(1) If $a, b, c \in R$, then $a < b \rightarrow a + c < b + c$.

(2) If $a, b \in R$ with $a > 0$ and $b > 0$, then $ab > 0$.

If R is a field, we call $(R, +, \cdot, <)$ an **ordered field**.

Note: If $a > 0$, we say that a is **positive**. If $a < 0$, we say that a is **negative**. If $a \geq 0$, we say that a is **nonnegative**. If $a \leq 0$, we say that a is **nonpositive**.

Example 5.8:

1. $(\mathbb{Z}, +, \cdot, <)$ is an ordered ring.

2. $(\mathbb{N}, +, \cdot, <)$ is **not** an ordered ring because $(\mathbb{N}, +, \cdot)$ is not a ring.

3. $(\mathbb{Q}, +, \cdot, <)$ and $(\mathbb{R}, +, \cdot, <)$ are ordered fields.

In general, we may just use the name of the underlying set for a whole structure when there is no danger of confusion. For example, we may refer to the ring R or the ordered field F instead of the ring $(R, +, \cdot)$ or the ordered field $(F, +, \cdot, <)$.

Ordered Field Fact 1: Let F be an ordered field. Then for all $x \in F^*$, $x \cdot x > 0$.

Analysis: There are two cases to consider:

Case 1: If $x > 0$, then by property (2) of an ordered field, $x \cdot x > 0$.

Case 2: If $x < 0$, then by property (1) of an ordered field, $x + (-x) < 0 + (-x)$, and so, $0 < -x$ (or equivalently, $-x > 0$). Therefore, by property (2) of an ordered field, $(-x)(-x) > 0$. Now, using Problems 59 and 60 from Problem Set 3, together with commutativity and associativity of multiplication in F, and the multiplicative identity property in F, we have

$$(-x)(-x) = (-1x)(-1x) = (-1)(-1)x \cdot x = 1(x \cdot x) = x \cdot x.$$

So, again we have $x \cdot x > 0$.

Ordered Field Fact 2: Let F be an ordered field and let $x \in F$ with $x > 0$. Then $\frac{1}{x} > 0$.

You will be asked to verify Ordered Field Fact 2 in Problems 67 through 71 below.

Why Isn't $\mathbb{Q}$ Enough?

At first glance, it would appear that the ordered field of rational numbers would be sufficient to solve all "real world" problems. However, a long time ago, a group of people called the Pythagoreans showed that this is not the case. The problem was first discovered when applying the now well-known Pythagorean Theorem.

Pythagorean Theorem: In a right triangle with legs of lengths a and b, and a hypotenuse of length c, $c^2 = a^2 + b^2$.

The picture to the right shows a right triangle. The vertical and horizontal segments (labeled a and b, respectively) are called the **legs** of the right triangle, and the side opposite the right angle (labeled c) is called the **hypotenuse** of the right triangle.

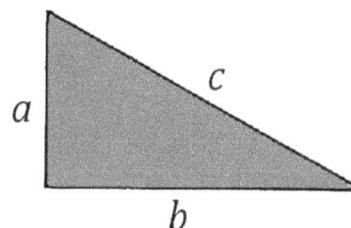

Example 5.9:

1. Consider a right triangle with legs of lengths $a = 3$ and $b = 4$, respectively. By the Pythagorean Thoerem, we have $c^2 = 3^2 + 4^2 = 9 + 16 = 25$, and so, $\boldsymbol{c = 5}$. That is, the hypotenuse of this right triangle has length **5**.

2. Consider a right triangle with a leg of length $a = 5$ and a hypotenuse of length $c = 13$. By the Pythagorean Theorem, we have $5^2 + b^2 = 13^2$, and so, $25 + b^2 = 169$. It follows that $b^2 = 169 - 25 = 144$, and so, $b = 12$. That is, the other leg has length **12**.

Exercise 5.10: Let a and b be the lengths of the legs of a right triangle and let c be the length of the hypotenuse of the right triangle. Find the third side length of the triangle, given the other two.

1. $a = 7, b = 24$ _____ $c = $ _____

2. $a = 12, c = 15$ _____ $b = $ _____

There are many ways to verify that the Pythagorean Theorem is true. Here, I will provide a simple geometric argument. The following two area formulas will be useful:

Area of a square with side length s: $A = s^2$

Area of a triangle with base b and height h: $A = \frac{1}{2}bh$

Verification of the Pythagorean Theorem: Consider a right triangle with legs of lengths a and b and hypotenuse of length c, as shown to the right.

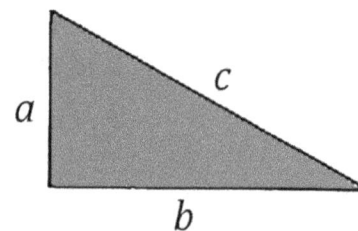

We draw 2 squares, each of side length $a + b$, by rearranging 4 copies of the given triangle in 2 different ways:

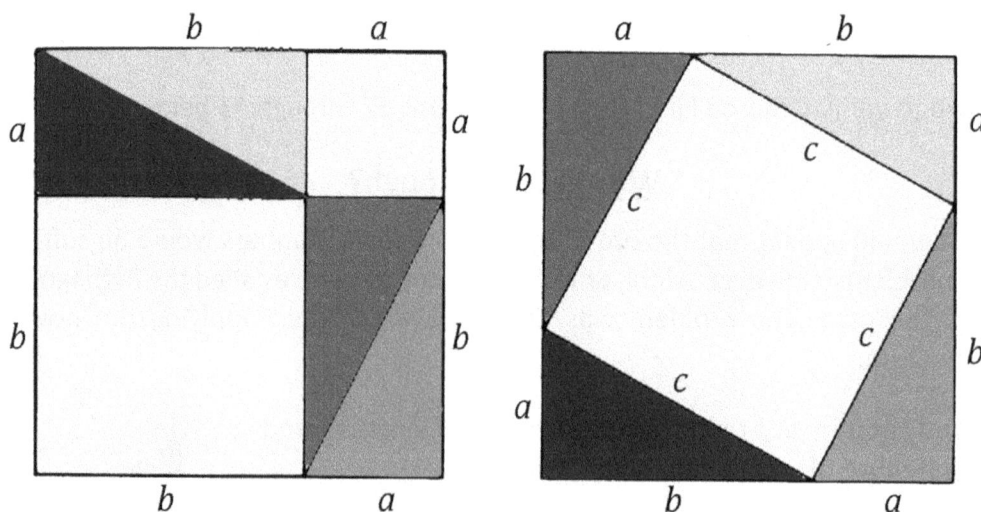

We can get the area of each of these squares by adding the areas of all the figures that comprise each square.

The square on the left consists of 4 copies of the given right triangle, a square of side length a and a square of side length b. It follows that the area of this square is $4 \cdot \frac{1}{2} ab + a^2 + b^2 = 2ab + a^2 + b^2$.

The square on the right consists of 4 copies of the given right triangle, and a square of side length c. It follows that the area of this square is $4 \cdot \frac{1}{2} ab + c^2 = 2ab + c^2$.

Since the areas of both squares of side length $a + b$ are equal (both areas are equal to $(a + b)^2$), $2ab + a^2 + b^2 = 2ab + c^2$. Cancelling $2ab$ from each side of this equation yields $a^2 + b^2 = c^2$.

Question: In a right triangle where both legs have length 1, what is the length of the hypotenuse?

Let's try to answer this question. If we let c be the length of the hypotenuse of the triangle, then by the Pythagorean Theorem, we have $c^2 = 1^2 + 1^2 = 1 + 1 = 2$. Since $c^2 = c \cdot c$, we need to find a number with the property that when you multiply that number by itself you get 2. The Pythagoreans showed that if we use only rational numbers, then no such number exists.

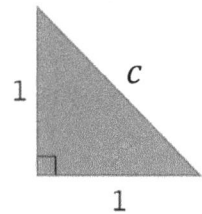

Rational Number Fact 1: There does **not** exist a rational number q such that $q^2 = 2$.

A verification of this fact will be explored in Problems 55 through 60 below.

Completeness

Let F be an ordered field and let S be a nonempty subset of F. We say that S is **bounded above** if there is $M \in F$ such that M is at least as big as every element of S. Each such number M is called an **upper bound** of S.

Example 5.11: In this example, $F = \mathbb{Q}$, the ordered field of rational numbers.

1. $S = \{1, 2, 3, 4, 5\}$ is bounded above in $\mathbb{Q}$.

   S has infinitely many upper bounds. For example, 5, 6, 12, and 1000 are upper bounds of S. In fact, any rational number greater than or equal to 5 is an upper bound of S.

   The number 5 is special in the sense that there are no upper bounds smaller than it. We say that 5 is the **least upper bound** of S.

   Notice that the least upper bound of S is inside the set S itself. This will always happen when the set S is finite. Specifically, if S is a finite set of rational numbers, then the least upper bound of S is simply the largest element of S.

2. $T = \{x \in \mathbb{Q} \mid -2 \leq x \leq 2\}$ is also bounded above in $\mathbb{Q}$.

   Any number greater than or equal to 2 is an upper bound of T and 2 is the least upper bound of T.

   Note that the least upper bound of T is in T.

3. $U = \{x \in \mathbb{Q} \mid x < -3\}$ is bounded above in $\mathbb{Q}$.

Any number greater than or equal to -3 is an upper bound of U and -3 is the least upper bound of U.

Note that in this case, the least upper bound of U is **not** in U.

4. $V = \{x \in \mathbb{Q} \mid x > 5\}$ is **not** bounded above in $\mathbb{Q}$.

To see this, observe that if M is a rational number greater than 5, then $M + 1 \in V$ and M is **not** as big as $M + 1$. So, M is **not** an upper bound of V.

5. $W = \{x \in \mathbb{Q} \mid x^2 < 2\}$ is bounded above by $2 \in \mathbb{Q}$.

To see this, note that if $x > 2$, then $x^2 > 4 \geq 2$, and therefore, $x \notin W$. Any number greater than 2 is also an upper bound.

Is 2 the least upper bound of V? It's not! For example, $\frac{3}{2}$ is also an upper bound. Indeed, if $x > \frac{3}{2}$, then $x^2 > \frac{9}{4} \geq 2$.

Does W have a least upper bound? A moment's thought might lead you to suspect that a least upper bound M would satisfy $M^2 = 2$. And it turns out that you are right! (Proving this, however, is quite difficult). This least upper bound M is not in the set W (if M were in W, then M^2 would be less than 2 and not equal to 2).

The big question is "Does M exist at all?"

By Rational Number Fact 1, M **does not** exist in $\mathbb{Q}$. So, W is an example of a set which is bounded above in $\mathbb{Q}$, but has no least upper bound in $\mathbb{Q}$.

So, if we want an ordered field F containing $\mathbb{Q}$ where M does exist, we can insist that F has the property that any set that is bounded above in F has a least upper bound in F. It turns out that $\mathbb{R}$ (the field of real numbers) is such an ordered field.

Exercise 5.12: Determine if each of the following sets is bounded above in $\mathbb{Q}$. If so, find the least upper bound of the set in $\mathbb{Q}$ if it exists.

1. $\mathbb{Z}$

2. $\mathbb{Z}^-$

3. $\{0, 3, 6, 9, 12\}$

4. $\left\{\frac{1}{n} \mid n \in \mathbb{Z}^+\right\}$

5. $\{x \in \mathbb{Q} \mid -7 < x < 135\}$

6. $\{x \in \mathbb{Q} \mid x > -500\}$

7. $\{x \in \mathbb{Q} \mid x \leq -500\}$

8. $\{x \in \mathbb{Q} \mid 0 < x^2 < 2\}$

Let F be an ordered field and let S be a nonempty subset of F. We say that S is **bounded below** if there is $M \in F$ such that M is no bigger than any element of S. Each such number M is called a **lower bound** of S.

Example 5.13: In this example, $F = \mathbb{Q}$, the ordered field of rational numbers.

1. $S = \{1, 2, 3, 4, 5\}$ is bounded below in $\mathbb{Q}$.

   S has infinitely many lower bounds. For example, $1, 0, -5$, and -1000 are lower bounds of S. In fact, any rational number less than or equal to 1 is a lower bound of S.

   The number 1 is special in the sense that there are no lower bounds larger than it. We say that 1 is the **greatest lower bound** of S.

   Notice that the greatest lower bound of S is inside the set S itself. This will always happen when the set S is finite. Specifically, if S is a finite set of rational numbers, then the greatest lower bound of S is simply the smallest element of S.

2. $T = \{x \in \mathbb{Q} \mid -2 \le x \le 2\}$ is also bounded below in $\mathbb{Q}$.

   Any number less than or equal to -2 is a lower bound of T and -2 is the greatest lower bound of T.

   Note that the greatest lower bound of T is in T.

3. $U = \{x \in \mathbb{Q} \mid x < -3\}$ is **not** bounded below in $\mathbb{Q}$.

   To see this, observe that if M is a rational number less than -3, then $M - 1 \in U$ and M is **not** as small as $M - 1$. So, M is **not** a lower bound of U.

4. $V = \{x \in \mathbb{Q} \mid x > 5\}$ is bounded below in $\mathbb{Q}$.

   Any number less than or equal to 5 is a lower bound of V and 5 is the greatest lower bound of V.

   Note that in this case, the greatest lower bound of V is **not** in V.

Exercise 5.14: Determine if each of the following sets is bounded below in $\mathbb{Q}$. If so, find the greatest lower bound of the set in $\mathbb{Q}$ if it exists.

1. $\mathbb{Z}$
2. $\mathbb{N}$
3. $\{0, 3, 6, 9, 12\}$
4. $\left\{ \frac{1}{n} \,\middle|\, n \in \mathbb{Z}^+ \right\}$
5. $\{x \in \mathbb{Q} \mid -7 < x < 135\}$
6. $\{x \in \mathbb{Q} \mid x > -500\}$
7. $\{x \in \mathbb{Q} \mid x \le -500\}$
8. $\{x \in \mathbb{Q} \mid 0 < x^2 < 2\}$

Let F be an ordered field and let S be a nonempty subset of F. We say that S is **bounded** if it is both bounded above and bounded below. Otherwise S is **unbounded**.

Example 5.15: In this example, $F = \mathbb{Q}$, the ordered field of rational numbers.

1. $S = \{1, 2, 3, 4, 5\}$ is bounded above by 5 and bounded below by 1. So, S is bounded.
2. $T = \{x \in \mathbb{Q} \mid -2 \le x \le 2\}$ is bounded above by 2 and bounded below by -2. So, T is bounded.
3. $U = \{x \in \mathbb{Q} \mid x < -3\}$ is **not** bounded below in $\mathbb{Q}$. So, U is unbounded.
4. $V = \{x \in \mathbb{Q} \mid x > 5\}$ is **not** bounded above in $\mathbb{Q}$. So, V is unbounded.
5. $W = \{x \in \mathbb{Q} \mid x^2 < 2\}$ is bounded above by 2 and bounded below by -2. So, W is bounded.

Exercise 5.16: Determine if each of the following sets is bounded or unbounded in $\mathbb{Q}$.

1. $\mathbb{Z}$
2. $\mathbb{N}$
3. $\{0, 3, 6, 9, 12\}$
4. $\left\{ \frac{1}{n} \,\middle|\, n \in \mathbb{Z}^+ \right\}$
5. $\{x \in \mathbb{Q} \mid -7 < x < 135\}$
6. $\{x \in \mathbb{Q} \mid x > -500\}$
7. $\{x \in \mathbb{Q} \mid x \le -500\}$
8. $\{x \in \mathbb{Q} \mid 0 < x^2 < 2\}$

An ordered field F has the **Completeness Property** if every nonempty subset of F that is bounded above in F has a least upper bound in F. In this case, we say that F is a **complete ordered field**.

Completeness Fact 1: There is exactly one complete ordered field (up to renaming the elements) containing the rational numbers. We call this field $\mathbb{R}$, the complete ordered field of real numbers.

We finish this lesson with two useful facts about $\mathbb{R}$.

Real Number Fact 1: The set of natural numbers is unbounded in the set of real numbers. In other words, if r is any real number, then we can find a natural number larger than r. This is known as the **Archimedean Property** of $\mathbb{R}$

Real Number Fact 2: Given any two distinct real numbers, we can find a rational number between them. This is known as the **Density Property** of $\mathbb{R}$.

A verification of Real Number Fact 1 will be explored in Problems 51 through 54 below.

The verification of Real Number Fact 2 is more difficult, and so, it is left as a Challenge Problem (Problem 72 below).

Problem Set 5

Full solutions to these problems are available for free download here:

www.SATPrepGet800.com/PMNR2ZX

LEVEL 1

Let S be the set of words in an English dictionary and define a relation $\prec$ on S by $x \prec y$ if x comes before y alphabetically. For example, alligator $\prec$ banana and dragon $\prec$ drainage.

1. Is $\prec$ reflexive on S?

2. Is $\prec$ symmetric on S?

3. Is $\prec$ transitive on S?

4. Is $\prec$ antireflexive on S?

5. Is $\prec$ antisymmetric on S?

6. Is $\prec$ trichotomous on S?

7. Does $\prec$ satisfy the comparability condition on S?

8. Is $(S, \prec)$ an ordered set?

Determine if each of the following is an ordered set.

9. $(\mathbb{Z}, >)$

10. $(\mathbb{Q}, <)$

11. $(\mathbb{Q}, \leq)$

12. $(\mathbb{R}, >)$

Determine if each of the following sets is bounded or unbounded in $\mathbb{Q}$.

13. $\mathbb{Q}$

14. $\mathbb{Z}^-$

15. $\{5, 10, 15, 20, 25, 30\}$

16. $\{1, 2, 3, \dots, 1000\}$

LEVEL 2

Let A be a set with at least two elements and consider the binary relation $\subset$ (proper subset) on $\mathcal{P}(A)$.

17. Is $\subset$ reflexive on $\mathcal{P}(A)$?

18. Is $\subset$ symmetric on $\mathcal{P}(A)$?

19. Is $\subset$ transitive on $\mathcal{P}(A)$?

20. Is $\subset$ antireflexive on $\mathcal{P}(A)$?

21. Is $\subset$ antisymmetric on $\mathcal{P}(A)$?

22. Is $\subset$ trichotomous on $\mathcal{P}(A)$?

23. Does $\subset$ satisfy the comparability condition on $\mathcal{P}(A)$?

24. Is $(\mathcal{P}(A), \subset)$ an ordered set?

Let a and b be the lengths of the legs of a right triangle and let c be the length of the hypotenuse of the right triangle. Find the third side length of the triangle, given the other two.

25. $a = 9, b = 12$

26. $a = 40, c = 41$

27. $a = 1, b = 2$

28. $b = 17, c = 26$

For each of the following sets, find the least upper bound in $\mathbb{Q}$ and greatest lower bound in $\mathbb{Q}$ if they exist.

29. $\{x \in \mathbb{Q} \mid -7 \leq x \leq 28\}$

30. $\{x \in \mathbb{Q} \mid x > 15\}$

31. $\mathbb{Q}^+$

32. $\{x \in \mathbb{Q} \mid -2 \leq x^2 \leq 2\}$

LEVEL 3

In Problems 33 through 36 we will show that there is no smallest positive real number. You may use Ordered Field Fact 2 for these problems. Let $x \in \mathbb{R}$ with $x > 0$.

33. Explain why $\frac{1}{2} > 0$.

34. Explain why $\frac{1}{2}x > 0$.

35. Explain why $x > \frac{1}{2}x$.

36. Use Problems 33 through 35 to explain why there is no smallest positive real number.

Let $a, r \in \mathbb{R}$ with $a \geq 0$.

37. Show that if $a = 0$, then a is less than every positive real number.

38. Show that if $a \neq 0$, then there is a positive real number less than a.

39. Explain why $a = 0$ if and only if a is less than every positive real number.

In Problems 40 through 42 we will use the Density Property of $\mathbb{R}$ (Real Number Fact 2) to show that given any two distinct real numbers, we can find an irrational number between them. Let $x, y \in \mathbb{R}$.

40. Suppose that $x < y$ and let c be a positive irrational number. Explain why there is a rational number q such that $\frac{x}{c} < q < \frac{y}{c}$.

41. Suppose that $x < y$ and let c be a positive irrational number. Explain why there is a **nonzero** rational number q such that $\frac{x}{c} < q < \frac{y}{c}$.

42. Suppose that $x < y$. Explain why there is an irrational number t such that $x < t < y$.

LEVEL 4

In Problems 43 through 46 we will use the Completeness Property of $\mathbb{R}$ to show that every nonempty set of real numbers that is bounded below has a greatest lower bound in $\mathbb{R}$. Let S be a nonempty set of real numbers that is bounded below and let $T = \{-x \mid x \in S\}$.

43. Let K be a lower bound for S. Explain why $-K$ is an upper bound for the set T.

44. Explain why T has a least upper bound M.

45. Let M be the least upper bound of T. Explain why $-M$ is a lower bound of S.

46. Let M be the least upper bound of T. Explain why $-M$ is a greatest lower bound of S.

In Problems 47 through 50 we will show that $\mathbb{Q}$ has the Archimedean Property. Let $\frac{a}{b}$ be a positive rational number.

47. Show that $(a + 1) - \frac{a}{b} = \frac{a(b-1)+b}{b}$.

48. Explain why $(a + 1) - \frac{a}{b} > 0$.

49. Explain why $a + 1 > \frac{a}{b}$.

50. Use Problems 47 through 49 to explain why $\mathbb{Q}$ has the Archimedean Property.

LEVEL 5

In Problems 51 through 54 we will show that $\mathbb{R}$ has the Archimedean Property.

51. Suppose that $\mathbb{N}$ is bounded from above in $\mathbb{R}$. Explain why this implies that $\mathbb{N}$ has a least upper bound in $\mathbb{R}$.

52. Let x be the least upper bound of $\mathbb{N}$ in $\mathbb{R}$ (assuming it exists). Explain why $x - 1$ is **not** an upper bound for $\mathbb{N}$.

53. Assuming that $x - 1$ is not an upper bound for $\mathbb{N}$, show that x is not an upper bound for $\mathbb{N}$.

54. Use Problems 51 through 53 to explain why $\mathbb{R}$ has the Archimedean Property.

In Problems 55 through 60 we will show that there is no rational number q such that $q^2 = 2$.

55. Let a be an integer such that a^2 is even. Explain why a must be even.

56. Let $\frac{c}{d}$ be a rational number. Explain why there is a rational number $\frac{a}{b}$ such that $\frac{a}{b} = \frac{c}{d}$ and either a is odd or b is odd (or both).

57. Let $\frac{a}{b}$ be a rational number such that $\left(\frac{a}{b}\right)^2 = 2$. Explain why a must be even.

58. Let $\frac{a}{b}$ be a rational number such that $\left(\frac{a}{b}\right)^2 = 2$. Explain why b must be even.

59. Explain why there cannot exist a rational number $\frac{a}{b}$ such that $\left(\frac{a}{b}\right)^2 = 2$ and either a is odd or b is odd (or both).

60. Explain why there does not exist a rational number $\frac{a}{b}$ such that $\left(\frac{a}{b}\right)^2 = 2$.

Suppose that $<$ is a strict linear ordering on $\mathbb{C}$ (the field of complex numbers) satisfying properties (1) and (2) of an ordered field.

61. Assume that $i > 0$. Use property (2) of an ordered field to explain why $-1 > 0$.

62. Assume that $i < 0$. Use property (1) of an ordered field to explain why $-i > 0$.

63. Assume that $-i > 0$. Use property (2) of an ordered field to explain why $-1 > 0$.

64. Assume that $-1 > 0$. Use property (1) of an ordered field to explain why $1 < 0$.

65. Assume that $-1 > 0$. Use property (2) of an ordered field to explain why $1 > 0$.

66. Use Problems 61 through 65 to explain why there is no way to turn $\mathbb{C}$ into an ordered field.

In Problems 67 through 71 we will verify Ordered Field Fact 2. Let F be an ordered field and let $x \in F$ with $x > 0$.

67. Explain why $\frac{1}{x} = x^{-1}$ exists and is nonzero.

68. Assume that $\frac{1}{x} < 0$. Use property (1) of an ordered field to explain why $-\frac{1}{x} > 0$.

69. Assume that $-\frac{1}{x} > 0$. Use property (2) of an ordered field to explain why $-1 > 0$.

70. Assume that $-1 > 0$. Use property (1) of an ordered field to explain why $1 < 0$.

71. Use Problems 67 through 70 to explain why $\frac{1}{x} > 0$.

CHALLENGE PROBLEMS

72. Verify that the Density Property of $\mathbb{R}$ (Real Number Fact 2) is true.

73. Show that there is a real number x such that $x^2 = 2$.

LESSON 6
TOPOLOGY

Intervals of Real Numbers

A set I of real numbers with at least two elements is called an **interval** if any real number that lies between two numbers in I is also in I.

Example 6.1:

1. The set $A = \{0, 1\}$ is **not** an interval. A consists of just the two real numbers 0 and 1. There are infinitely many real numbers between 0 and 1. For example, the real number $\frac{1}{2}$ is between 0 and 1.

2. The set $B = \{x \in \mathbb{R} \mid 0 < x < 1\}$ is an example of an **open interval**. This set consists of all real numbers between 0 and 1, exclusive (0 and 1 are excluded). We will usually write the set B using the **interval notation** $B = (0, 1)$.

3. The set $C = \{x \in \mathbb{R} \mid 0 \leq x \leq 1\}$ is an example of a **closed interval**. This set consists of all real numbers between 0 and 1, inclusive (0 and 1 are included). We will usually write the set C using the **interval notation** $C = [0, 1]$.

4. The set $D = \{x \in \mathbb{R} \mid 0 \leq x < 1\}$ is an example of a **half-open interval**. This set consists of all real numbers between 0 and 1, including 0, but excluding 1. We will usually write the set D using the **interval notation** $D = [0, 1)$.

5. The set $E = \{x \in \mathbb{R} \mid x > 1\}$ is an example of an **infinite open interval**. This set consists of all real numbers greater than 1. We will usually write the set E using the **interval notation** $E = (1, \infty)$. The symbol "∞" is pronounced "**infinity**." It is **not** a number, but rather a symbol indicating that the set E has no upper bound.

6. $\mathbb{R}$ is an interval. This follows trivially from the definition. After all, any real number that lies between two real numbers is a real number.

Note: Observe how the "interval notation" mentioned in 2, 3, and 4 above uses parentheses "( )" to indicate that endpoints are **not** included in the set and brackets "[ ]" to indicate that endpoints are included in the set.

Exercise 6.2: Determine if each of the following sets of real numbers is an interval.

1. $F = \{x \in \mathbb{R} \mid 0 < x \leq 1\}$ _____

2. $\mathbb{Z}$ _____

3. $G = \left\{ \frac{1}{n} \mid n \in \mathbb{Z}^+ \right\}$ _____

4. $H = \left\{ x \in \mathbb{R} \mid \frac{1}{5} < x < \frac{1}{4} \right\}$ _____

5. $I = \{x \in \mathbb{R} \mid x \leq 2.7\}$ _____

6. $\mathbb{Q}$ _____

When we are thinking of $\mathbb{R}$ as an interval, we sometimes use the notation $(-\infty, \infty)$ and refer to this as **the real line**. The following picture gives the standard geometric interpretation of the real line.

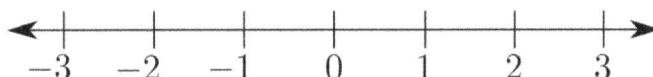

In addition to the real line, there are 8 other types of intervals.

Open Interval: $(a, b) = \{x \in \mathbb{R} \mid a < x < b\}$

Closed Interval: $[a, b] = \{x \in \mathbb{R} \mid a \leq x \leq b\}$

Half-open Intervals: $(a, b] = \{x \in \mathbb{R} \mid a < x \leq b\}$ $[a, b) = \{x \in \mathbb{R} \mid a \leq x < b\}$

Infinite Open Intervals: $(a, \infty) = \{x \in \mathbb{R} \mid x > a\}$ $(-\infty, b) = \{x \in \mathbb{R} \mid x < b\}$

Infinite Closed Intervals: $[a, \infty) = \{x \in \mathbb{R} \mid x \geq a\}$ $(-\infty, b] = \{x \in \mathbb{R} \mid x \leq b\}$

Notes: (1) Each of the nine types of sets above (including the real line) satisfies the definition of being an interval. Conversely, every interval has one of these nine forms.

(2) The first four intervals above (the open, closed, and two half-open intervals) are **bounded**. They are each bounded below by a and bounded above by b. In fact, for each of these intervals, a is the **greatest lower bound** and b is the **least upper bound**.

Example 6.3: The half-open interval $(-2, 1] = \{x \in \mathbb{R} \mid -2 < x \leq 1\}$ has the following graph:

Note: The left parenthesis appearing at -2 indicates that -2 is **not** included in the set, whereas the right bracket appearing at 1 indicates that 1 is included in the set.

Exercise 6.4: Sketch the graph of the infinite open interval $(0, \infty) = \{x \in \mathbb{R} \mid x > 0\}$.

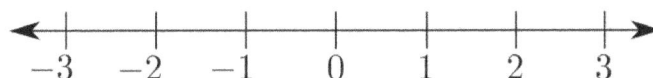

More Set Operations

In Lesson 2 we saw how to take the union, intersection, difference, and symmetric difference of two sets. We will now review the definitions from that lesson and apply those definitions to intervals. We will then generalize the definitions of union and intersection to more than two sets.

The **union** of the sets A and B, written $A \cup B$, is the set of elements that are in A or B (or both).

$$A \cup B = \{x \mid x \in A \text{ or } x \in B\}$$

The **intersection** of A and B, written $A \cap B$, is the set of elements that are simultaneously in A and B.

$$A \cap B = \{x \mid x \in A \text{ and } x \in B\}$$

The **difference** $A \setminus B$ is the set of elements that are in A and not in B.

$$A \setminus B = \{x \mid x \in A \text{ and } x \notin B\}$$

The **symmetric difference** between A and B, written $A \, \Delta \, B$, is the set of elements that are in A or B, but not both.

$$A \, \Delta \, B = (A \setminus B) \cup (B \setminus A)$$

The following Venn diagrams can be useful for visualizing these operations. As usual, U is some "universal" set that contains both A and B.

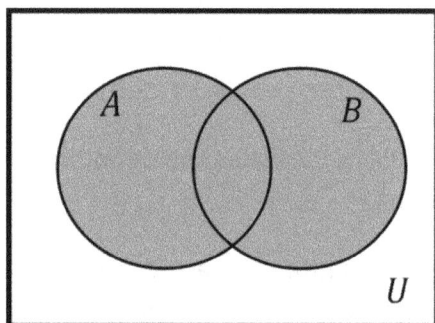

$A \cup B$

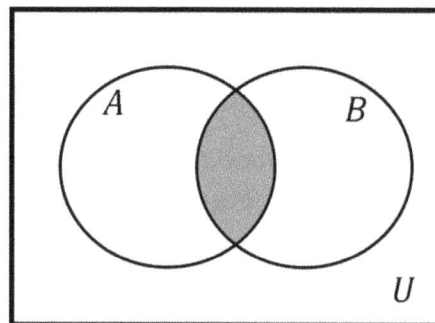

$A \cap B$

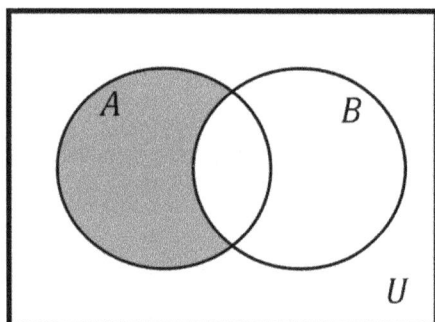

$A \setminus B$

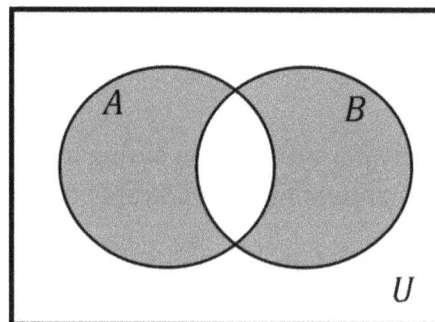

$A \, \Delta \, B$

Example 6.5: Let $A = (-2, 1]$ and $B = (0, \infty)$. We have

1. $A \cup B = (-2, \infty)$
2. $A \cap B = (0, 1]$
3. $A \setminus B = (-2, 0]$
4. $B \setminus A = (1, \infty)$
5. $A \, \Delta \, B = (-2, 0] \cup (1, \infty)$

Note: If you have trouble seeing how to compute these, it may be helpful to draw the graphs of A and B lined up vertically, and then draw vertical lines through the endpoints of each interval.

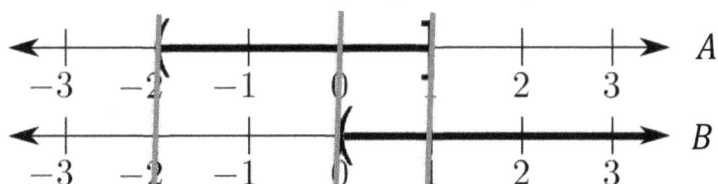

The results follow easily by combining these graphs into a single graph using the vertical lines as guides. For example, let's look at $A \cap B$ in detail. We're looking for all numbers that are in both A and B. The two rightmost vertical lines drawn passing through the two graphs above isolate all those numbers nicely. We see that all numbers between 0 and 1 are in the intersection. We should then think about the two endpoints 0 and 1 separately. $0 \notin B$ and therefore, 0 cannot be in the intersection of A and B. On the other hand, $1 \in A$ and $1 \in B$. Therefore, $1 \in A \cap B$. So, we see that $A \cap B = (0, 1]$.

Exercise 6.6: Let $C = (-\infty, 2]$, $D = (-1, 3]$. Compute each of the following:

1. $C \cup D$ _____

2. $C \cap D$ _____

3. $C \setminus D$ _____

4. $D \setminus C$ _____

5. $C \mathbin{\Delta} D$ _____

In topology, we will often want to look at unions and intersections of more than two sets. Therefore, we make the following more general definitions.

Let $\boldsymbol{X}$ be a nonempty set of sets.

$$\bigcup \boldsymbol{X} = \{y \mid \text{there is } Y \in X \text{ with } y \in Y\} \qquad \text{and} \qquad \bigcap \boldsymbol{X} = \{y \mid \text{for all } Y \in X, y \in Y\}.$$

If you're having trouble understanding what these definitions are saying, you're not alone. The notation probably looks confusing, but the ideas behind these definitions are very simple. You have a whole bunch of sets (possibly infinitely many). To take the union of all these sets, you simply throw all the elements together into one big set. To take the intersection of all these sets, you take only the elements that are in every single one of those sets.

Example 6.7:

1. Let A and B be sets and let $\boldsymbol{X} = \{A, B\}$. Then

$$\bigcup \boldsymbol{X} = \{y \mid \text{there is } Y \in X \text{ with } y \in Y\} = \{y \mid y \in A \text{ or } y \in B\} = A \cup B.$$
$$\bigcap \boldsymbol{X} = \{y \mid \text{for all } Y \in X, y \in Y\} = \{y \mid y \in A \text{ and } y \in B\} = A \cap B.$$

Another way to write this is $\bigcup\{A, B\} = A \cup B$ and $\bigcap\{A, B\} = A \cap B$.

2. Let A, B, and C be sets, and let $X = \{A, B, C\}$. Then

$$\cup X = \{y \mid \text{there is } Y \in X \text{ with } y \in Y\} = \{y \mid y \in A, y \in B, \text{or } y \in C\} = A \cup B \cup C.$$
$$\cap X = \{y \mid \text{for all } Y \in X, y \in Y\} = \{y \mid y \in A, y \in B, \text{and } y \in C\} = A \cap B \cap C.$$

Another way to write this is $\cup\{A, B, C\} = A \cup B \cup C$ and $\cap\{A, B, C\} = A \cap B \cap C$.

3. Let $X = \{[0, r) \mid r \in \mathbb{R}^+\}$. Then we have $\cup X = [0, \infty)$ and $\cap X = \{0\}$.

Another way to write this is $\cup\{[0, r) \mid r \in \mathbb{R}^+\} = [0, \infty)$ and $\cap\{[0, r) \mid r \in \mathbb{R}^+\} = \{0\}$.

Exercise 6.8: For each of the following, compute $\cup X$ and $\cap X$.

1. $X = \{\{a, b, c\}, \{b, c, d, e\}, \{a, b, x, y, z\}\}$ $\cup X = $ _____ $\cap X = $ _____

2. $X = \{(0, 4), [1, 5), [-17, 3), (2, \infty)\}$ $\cup X = $ _____ $\cap X = $ _____

3. $X = \{\{-n, \ldots, -3, -2, -1, 0, 1, 2, 3, 4, \ldots, n\} \mid n \in \mathbb{N}\}$ $\cup X = $ _____ $\cap X = $ _____

Open Sets in $\mathbb{R}$

A subset X of $\mathbb{R}$ is said to be **open** in $\mathbb{R}$ if for every real number $x \in X$, there is an open interval (a, b) with $x \in (a, b)$ and $(a, b) \subseteq X$.

In words, a set is open in $\mathbb{R}$ if every number in the set has "some space" on both sides of that number inside the set. If you think of each point in the set as an animal, then each animal in the set should be able to move a little to the left and a little to the right without ever leaving the set. Another way to think of this is that no number is on "the edge" or "the boundary" of the set, about to fall out of it.

Example 6.9:

1. The open interval $(0, 1)$ is open in $\mathbb{R}$. Indeed, if $x \in (0, 1)$, then $(0, 1)$ itself is an open interval with $x \in (0, 1)$ and $(0, 1) \subseteq (0, 1)$.

2. In fact, every bounded open interval (a, b) is open in $\mathbb{R}$. Indeed, if $x \in (a, b)$, then (a, b) itself is an open interval with $x \in (a, b)$ and $(a, b) \subseteq (a, b)$. For example, $(5, 7)$ and $\left(-\sqrt{2}, \frac{3}{5}\right)$ are open in $\mathbb{R}$.

3. All open intervals are open in $\mathbb{R}$, even unbounded ones. For example, $(-2, \infty)$, $(-\infty, 5)$, and $(-\infty, \infty)$ are all open in $\mathbb{R}$.

4. $(0, 1]$ is **not** open in $\mathbb{R}$ because the "boundary point" 1 is included in the set. If (a, b) is any open interval containing 1, then $(a, b) \nsubseteq (0, 1]$ because there are numbers greater than 1 inside (a, b).

5. We can use reasoning similar to that used in part 4 above to see that all half-open intervals and closed intervals are **not** open in $\mathbb{R}$.

Exercise 6.10: Determine if each of the following sets is open in $\mathbb{R}$.

1. $(-3, 8)$ _____

2. $[2, 9]$ _____

3. $(-\infty, -6)$ _____

4. $[0, \infty)$ _____

5. $\mathbb{R}$ _____

6. $\emptyset$ _____

The following basic facts about open sets are useful.

Open Set Fact 1: An arbitrary union of open sets in $\mathbb{R}$ is open in $\mathbb{R}$.

Open Set Fact 2: Every nonempty open set in $\mathbb{R}$ can be expressed as a union of bounded open intervals.

Open Set Fact 3: A finite intersection of open sets in $\mathbb{R}$ is open in $\mathbb{R}$.

Example 6.11:

1. $(-5, 2)$ is open in $\mathbb{R}$ by part 2 of Example 6.9 and $(7, \infty)$ is open in $\mathbb{R}$ by part 3 of Example 6.9. Therefore, by Open Set Fact 1, $(-5, 2) \cup (7, \infty)$ is also open in $\mathbb{R}$.

2. $(1, 2) \cup (2, 3) \cup (3, 4) \cup (4, \infty)$ is open in $\mathbb{R}$ by Open Set Fact 1.

3. $\mathbb{R} \setminus \mathbb{Z}$ is open because it is a union of open intervals. It looks like this:

$$\cdots (-2, -1) \cup (-1, 0) \cup (0, 1) \cup (1, 2) \cup \cdots$$

$\mathbb{R} \setminus \mathbb{Z}$ can also be written as

$$\bigcup \{(n, n+1) \mid n \in \mathbb{Z}\} \qquad \text{or} \qquad \bigcup_{n \in \mathbb{Z}} (n, n+1)$$

4. If we take the union of all intervals of the form $\left(\frac{1}{n+1}, \frac{1}{n}\right)$ for positive integers n, we get an open set. We can visualize this open set as follows:

$$\bigcup \left\{ \left(\frac{1}{n+1}, \frac{1}{n}\right) \,\Big|\, n \in \mathbb{Z}^+ \right\} = \cdots \cup \left(\frac{1}{5}, \frac{1}{4}\right) \cup \left(\frac{1}{4}, \frac{1}{3}\right) \cup \left(\frac{1}{3}, \frac{1}{2}\right) \cup \left(\frac{1}{2}, 1\right)$$

Exercise 6.12: Determine if each of the following sets is open in $\mathbb{R}$.

1. $(-\infty, 3) \cup (5, 11) \cup (17, \infty)$ _____

2. $\mathbb{Q}$ _____

3. $\mathbb{R} \setminus \mathbb{Q}$ _____

4. $\bigcup \left\{ \left(n, n+\frac{1}{2}\right) \,\Big|\, n \in \mathbb{Z} \right\}$ _____

5. $(-\infty, 3) \cap (0, 7)$ _____

6. $\bigcap \left\{ \left(0, 1+\frac{1}{n}\right) \,\Big|\, n \in \mathbb{Z} \right\}$ _____

Closed Sets in $\mathbb{R}$

A subset X of $\mathbb{R}$ is said to be **closed** in $\mathbb{R}$ if $\mathbb{R} \setminus X$ is open in $\mathbb{R}$.

Note: $\mathbb{R} \setminus X$ is called the **complement** of X in $\mathbb{R}$, or simply the complement of X. It consists of all real numbers **not** in X.

Example 6.13:

1. The closed interval $[0, 1]$ is closed in $\mathbb{R}$. Its complement in $\mathbb{R}$ is $\mathbb{R} \setminus [0, 1] = (-\infty, 0) \cup (1, \infty)$. This is a union of open intervals, which is open in $\mathbb{R}$.

2. In fact, every bounded closed interval $[a, b]$ is closed in $\mathbb{R}$. Indeed, the complement of $[a, b]$ in $\mathbb{R}$ is $\mathbb{R} \setminus [a, b] = (-\infty, a) \cup (b, \infty)$. This is a union of open intervals, which is open in $\mathbb{R}$. For example, $[5, 7]$ and $\left[-\sqrt{2}, \frac{3}{5}\right]$ are closed in $\mathbb{R}$.

3. All closed intervals are open in $\mathbb{R}$, even unbounded ones. For example, $[3, \infty)$ is closed in $\mathbb{R}$ because $\mathbb{R} \setminus [3, \infty) = (-\infty, 3)$, which is open in $\mathbb{R}$.

4. Half-open intervals are neither open nor closed in $\mathbb{R}$. For example, we saw in part 4 of Example 6.9 that $(0, 1]$ is **not** open in $\mathbb{R}$. We see that $(0, 1]$ is not closed in $\mathbb{R}$ by observing that $\mathbb{R} \setminus (0, 1] = (-\infty, 0] \cup (1, \infty)$, which is not open in $\mathbb{R}$.

5. If $a \in \mathbb{R}$, then the set $\{a\}$ consisting of just one real number is closed in $\mathbb{R}$ because it is the complement of the open set $(-\infty, a) \cup (a, \infty)$.

Exercise 6.14: Determine if each of the following sets is closed in $\mathbb{R}$.

1. $(-3, 8)$ _____

2. $[2, 9]$ _____

3. $(-\infty, -6)$ _____

4. $[0, \infty)$ _____

5. $\mathbb{R}$ _____

6. $\emptyset$ _____

The following basic facts about closed sets are useful.

Closed Set Fact 1: An arbitrary intersection of closed sets in $\mathbb{R}$ is closed in $\mathbb{R}$.

Closed Set Fact 2: A finite union of closed sets in $\mathbb{R}$ is closed in $\mathbb{R}$.

Example 6.15:

1. $[-5, 2]$ is closed in $\mathbb{R}$ by part 2 of Example 6.13 and $[7, \infty)$ is closed in $\mathbb{R}$ by part 3 of Example 6.13. Therefore, by Closed Set Fact 2, $[-5, 2] \cup [7, \infty)$ is also closed in $\mathbb{R}$.

2. $[1, 2] \cup [2, 3] \cup [3, 4] \cup [4, \infty)$ is closed in $\mathbb{R}$ by Closed Set Fact 2.

3. Since the union of finitely many closed sets is closed and a set containing a single point is closed (by part 5 of Example 6.13), it follows that every finite set is closed.

Problem Set 6

LEVEL 1

Determine if each of the following sets is an interval.

1. $A = \{x \in \mathbb{R} \mid 12 \leq x \leq 15\}$

2. $B = \{x \in \mathbb{R} \mid x < -103\}$

3. $C = \{x \in \mathbb{Q} \mid x < -103\}$

4. $D = \mathbb{Q}^-$

5. $E = \mathbb{R}^+$

6. $F = \{x \in \mathbb{R} \mid x \geq -16\}$

7. $G = \{x \in \mathbb{R} \mid 0 \leq x < 999\}$

8. $\mathbb{R} \setminus \{0\}$

Sketch the graph of each of the following:

9. $\mathbb{R}$

10. $\mathbb{R}^+$

11. $\{-1, 1\}$

12. $(-1, 1)$

13. $[-1, \infty)$

14. $\mathbb{N}$

15. $\mathbb{Z}$

16. $(-\infty, -1)$

17. $(1, 2]$

Let $A = (-17, 6)$ and $B = (-1, 17)$. Compute each of the following:

18. $A \cup B$

19. $A \cap B$

20. $A \setminus B$

21. $B \setminus A$

22. $A \Delta B$

Let $A = [14, \infty)$ and $B = (-\infty, 15)$. Compute each of the following:

23. $A \cup B$

24. $A \cap B$

25. $A \setminus B$

26. $B \setminus A$

27. $A \Delta B$

For each of the following, compute $\cup X$ and $\cap X$.

28. $X = \{\{0, 1, 2\}, \{1, 2, 3\}, \{2, 3, 4\}\}$

29. $X = \{\mathbb{N}, \mathbb{Z}, \mathbb{Q}, \mathbb{R}\}$

30. $X = \{(0, 10), (1, 11), (2, 12), (3, 13), (4, 14)\}$

31. $X = \{(-\infty, 20), [-5, 17), (4, 100]\}$

32. $X = \{(0, 1], (1, 2], (2, 3], (3, 4]\}$

Determine if each of the following sets is open, closed, both, or neither.

33. $(5, 11]$

34. $\{x \in \mathbb{R} \mid x < -103\}$

35. $(12, 17) \cup (26, \infty)$

36. $(0, 5) \cap (2, 6]$

37. $(0, 5) \cap [2, 6)$

38. $(0, 5] \cap [2, \infty)$

39. $(0, 1) \cap (1, 2)$

Let S be a set of real numbers. A real number x is called an **accumulation point** of S if every open interval containing x contains at least one point of S different from x. Find the accumulation points of each of the following sets:

40. $\left\{1, \frac{1}{2}, \frac{1}{3}, \ldots\right\}$

41. $[0, 1)$

42. $\mathbb{Z}$

43. $\mathbb{Q}$

LEVEL 4

For each of the following, compute $\cup X$ and $\cap X$.

44. $X = \{(0, q] \mid q \in \mathbb{Q}^+\}$

45. $X = \left\{\left(\frac{1}{n}, 1\right) \mid n \in \mathbb{Z}^+\right\}$

46. $X = \{(n, n + 3) \mid n \in \mathbb{Z}\}$

47. $X = \left\{\left(0, 1 + \frac{1}{n}\right) \mid n \in \mathbb{Z}^+\right\}$

48. $X = \left\{\left(-\infty, \frac{1}{n}\right] \mid n \in \mathbb{Z}^+\right\}$

123

Determine if each of the following sets is open, closed, both, or neither.

49. $\mathbb{Q}^+$

50. $\mathbb{Z}$

51. $\mathbb{R} \setminus \mathbb{N}$

52. $\cup \left\{ \left(0, 1 + \frac{1}{n} \right) \mid n \in \mathbb{Z}^+ \right\}$

53. $\cap \left\{ \left[0, 1 - \frac{1}{n} \right] \mid n \in \mathbb{Z}^+ \right\}$

If X is a nonempty set of sets, we say that X is **disjoint** if $\cap X = \emptyset$. We say that X is **pairwise disjoint** if for all $A, B \in X$ with $A \neq B$, A and B are disjoint. For each of the following, determine if X is disjoint, pairwise disjoint, both, or neither.

54. $\{ (n, n+1) \mid n \in \mathbb{Z} \}$

55. $\{ (n, n+1] \mid n \in \mathbb{Z} \}$

56. $\left\{ \left(\frac{1}{n+1}, \frac{1}{n} \right) \mid n \in \mathbb{Z}^+ \right\}$

57. $\{ \mathbb{Q} \}$

58. $\left\{ \mathbb{Q}^-, 2\mathbb{N}, \{ n \in \mathbb{N} \mid n \text{ is a prime number greater than } 2 \} \right\}$

LEVEL 5

Let X be a nonempty set of sets. Verify each of the following:

59. For all $A \in X$, $A \subseteq \cup X$.

60. For all $A \in X$, $\cap X \subseteq A$.

Let A be a set and let $\boldsymbol{X}$ be a nonempty set of sets. Verify each of the following:

61. $A \cap \cup \boldsymbol{X} = \cup\{A \cap B \mid B \in \boldsymbol{X}\}$.

62. $A \cup \cap \boldsymbol{X} = \cap\{A \cup B \mid B \in \boldsymbol{X}\}$.

63. $A \setminus \cup \boldsymbol{X} = \cap\{A \setminus B \mid B \in \boldsymbol{X}\}$.

64. $A \setminus \cap \boldsymbol{X} = \cup\{A \setminus B \mid B \in \boldsymbol{X}\}$.

Verify each of the following:

65. An arbitrary union of open sets in $\mathbb{R}$ is an open set in $\mathbb{R}$.

66. An arbitrary intersection of closed sets in $\mathbb{R}$ is closed in $\mathbb{R}$.

CHALLENGE PROBLEMS

67. Let S be a subset of $\mathbb{R}$. Recall that a real number x is called an **accumulation point** of S if every open interval containing x contains at least one point of S different from x. Show that a subset S of $\mathbb{R}$ is closed in $\mathbb{R}$ if and only if S contains each of its accumulation points.

68. Recall that a nonempty set $\boldsymbol{X}$ of sets is **pairwise disjoint** if for all $A, B \in \boldsymbol{X}$ with $A \neq B$, A and B are disjoint. Prove that every nonempty open set of real numbers can be expressed as a union of pairwise disjoint open intervals.

125

LESSON 7
COMPLEX ANALYSIS

The Complex Field

The **standard form of a complex number** is $a + bi$, where a and b are real numbers. So, the set of complex numbers is $\mathbb{C} = \{a + bi \mid a, b \in \mathbb{R}\}$.

In addition to visualizing the complex number $a + bi$ as the point (a, b) in the Complex Plane (see Lesson 2), we can also visualize $a + bi$ as a directed line segment (or **vector**) starting at the origin and ending at the point (a, b). Three examples are shown to the right, namely $1 + 2i$, $2 = 2 + 0i$, and $i = 0 + 1i$.

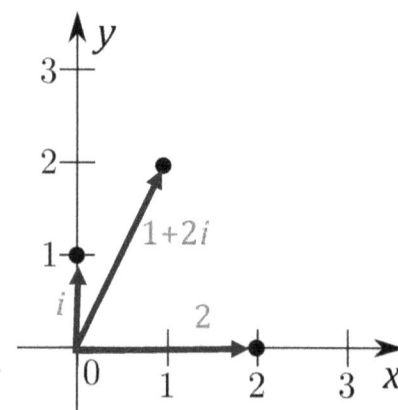

If $z = a + bi$ is a complex number, we call a the **real part** of z and b the **imaginary part** of z, and we write $a = \operatorname{Re} z$ and $b = \operatorname{Im} z$.

Two complex numbers are **equal** if and only if they have the same real part and the same imaginary part. In other words,

$$a + bi = c + di \text{ if and only if } a = c \text{ and } b = d.$$

We add two complex numbers by simply adding their real parts and adding their imaginary parts. So,

$$(a + bi) + (c + di) = (a + c) + (b + d)i.$$

As a point, this sum is $(a + c, b + d)$. We can visualize this sum as the vector starting at the origin that is the diagonal of the parallelogram formed from the vectors $a + bi$ and $c + di$. Here is an example showing that $(1 + 2i) + (-3 + i) = -2 + 3i$.

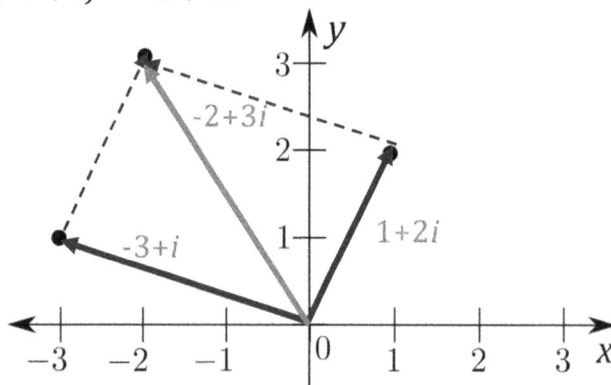

Example 7.1:

1. $(5 + 3i) + (2 + i) = (5 + 2) + (3 + 1)i = \mathbf{7 + 4i}$

2. $(-7 + 6i) + (4 + (-5)i) = (-7 + 4) + \big(6 + (-5)\big)i = \mathbf{-3 + i}$

3. $(2 - 5i) + (3 - 6i) = (2 + 3) + \big(-5 + (-6)\big)i = \mathbf{5 - 11i}$

126

Exercise 7.2: Compute each of the following sums:

1. $(1 + i) + (1 - i)$ = _____

2. $(-3 - 2i) + (-5 + i)$ = _____

3. $5i + (9 - 3i)$ = _____

The definition for multiplying two complex numbers is a bit more complicated:

$$(a + bi)(c + di) = (ac - bd) + (ad + bc)i.$$

Example 7.3:

1. $(5 + 3i)(2 + i) = (5 \cdot 2 - 3 \cdot 1) + (5 \cdot 1 + 3 \cdot 2)i = (10 - 3) + (5 + 6)i = \mathbf{7 + 11i}$

2. $(-7 + 6i)(4 + (-5)i) = (-7 \cdot 4 - 6(-5)) + ((-7)(-5) + 6 \cdot 4)i$

   $= (-28 + 30) + (35 + 24)i = \mathbf{2 + 59i}$

3. $(2 - 5i)(3 - 6i) = (2 \cdot 3 - (-5)(-6)) + (2(-6) + (-5) \cdot 3)i$

   $= (6 - 30) + (-12 - 15)i = \mathbf{-24 - 27i}$

Exercise 7.4: Compute each of the following products:

1. $(1 + i)(1 - i)$ = _____

2. $(-3 - 2i)(-5 + i)$ = _____

3. $5i(9 - 3i)$ = _____

Notes: (1) If $b = 0$, then we call $a + bi = a + 0i = a$ a **real number**. Note that when we add or multiply two real numbers, we always get another real number.

$$(a + 0i) + (b + 0i) = (a + b) + (0 + 0)i = (a + b) + 0i = a + b.$$

$$(a + 0i)(b + 0i) = (ab - 0 \cdot 0) + (a \cdot 0 + 0b)i = (ab - 0) + (0 + 0)i = ab + 0i = ab.$$

(2) If $a = 0$, then we call $a + bi = 0 + bi = bi$ a **pure imaginary number**.

(3) $i^2 = -1$. To see this, note that $i^2 = i \cdot i = (0 + 1i)(0 + 1i)$, and we have

$$(0 + 1i)(0 + 1i) = (0 \cdot 0 - 1 \cdot 1) + (0 \cdot 1 + 1 \cdot 0)i = (0 - 1) + (0 + 0)i = -1 + 0i = -1.$$

(4) The definition of the product of two complex numbers is motivated by how multiplication should behave in a field, together with replacing i^2 by -1. If we were to naïvely multiply the two complex numbers, we would have

$$(a + bi)(c + di) = (a + bi)c + (a + bi)(di) = ac + bci + adi + bdi^2$$
$$= ac + bci + adi + bd(-1) = ac + (bc + ad)i - bd = (ac - bd) + (ad + bc)i.$$

The dedicated reader may want to make a note of which field properties were used during this computation.

Those familiar with the mnemonic FOIL may notice that "FOILing" will always work to produce the product of two complex numbers, provided we replace i^2 by -1 and simplify.

With the definitions we just made for addition and multiplication, we get $(\mathbb{C}, +, \cdot)$, the **field of complex numbers**. In problems 67 through 71 in Problem Set 3, you were asked to verify that $(\mathbb{C}, +, \cdot)$ does in fact form a field.

We review a few items of importance here:

- The identity for addition is $0 = 0 + 0i$.

- The identity for multiplication is $1 = 1 + 0i$

- The additive inverse of $z = a + bi$ is $-z = -(a + bi) = -a - bi$.

- The multiplicative inverse of $z = a + bi$ is $z^{-1} = \frac{a}{a^2+b^2} - \frac{b}{a^2+b^2}i$.

Technical remark: By Note 1 following Exercise 7.4, we see that $(\mathbb{R}, +, \cdot)$ is a **subfield** of $(\mathbb{C}, +, \cdot)$. That is, $\mathbb{R} \subseteq \mathbb{C}$ and $(\mathbb{R}, +, \cdot)$ is a field with respect to the field operations of $(\mathbb{C}, +, \cdot)$ (In other words, we don't need to "change" the definition of addition or multiplication to get the appropriate operations in $\mathbb{R}$—the operations are already behaving correctly).

Exercise 7.5: Find the additive and multiplicative inverses of each of the following complex numbers:

1. $1 + i$
2. $2 - 3i$
3. $5i$
4. 7
5. $\sqrt{2} - \sqrt{5}i$

Subtraction: If $z, w \in \mathbb{C}$, with $z = a + bi$ and $w = c + di$, then we define the **difference** $z - w$ by
$$z - w = z + (-w) = (a + bi) + (-c - di) = (a - c) + (b - d)i.$$

As a point, this difference is $(a - c, b - d)$.

Example 7.6:

1. $(5 + 3i) - (2 + i) = (5 - 2) + (3 - 1)i = \mathbf{3 + 2i}$
2. $(-7 + 6i) - (4 + (-5)i) = (-7 - 4) + (6 - (-5))i = \mathbf{-11 + 11i}$
3. $(2 - 5i) - (3 - 6i) = (2 - 3) + (-5 - (-6))i = \mathbf{-1 + i}$

Exercise 7.7: Compute each of the following differences:

1. $(1 + i) - (1 - i)$ =   _____
2. $(-3 - 2i) - (-5 + i)$ =   _____
3. $5i - (9 - 3i)$ =   _____

Below is an example illustrating how subtraction works using the following computation:

$$(1 + 2i) - (2 - i) = -1 + 3i.$$

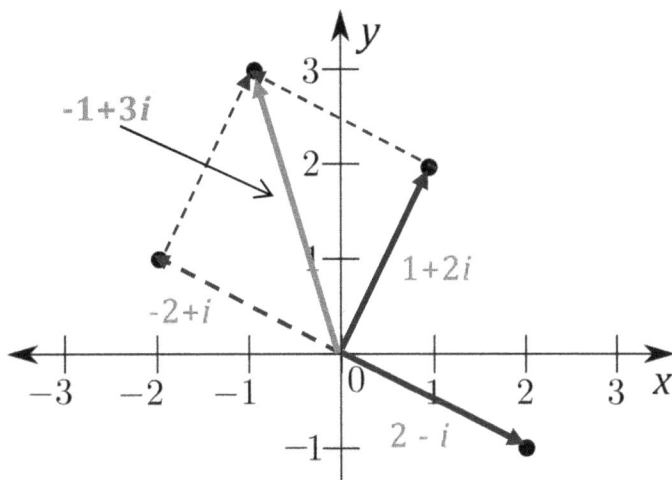

Observe how we first replaced $2 - i$ by $-2 + i$ so that we could change the subtraction problem to the addition problem: $(1 + 2i) + (-2 + i)$. We then formed a parallelogram using $1 + 2i$ and $-2 + i$ as edges, and finally, drew the diagonal of that parallelogram to see the result.

Note: Geometrically, we can translate the vector $-1 + 3i$ so that its initial point coincides with the terminal point of the vector $2 - i$. When we do this, the terminal point of $-1 + 3i$ coincides with the terminal point of the vector $1 + 2i$. See the figure below.

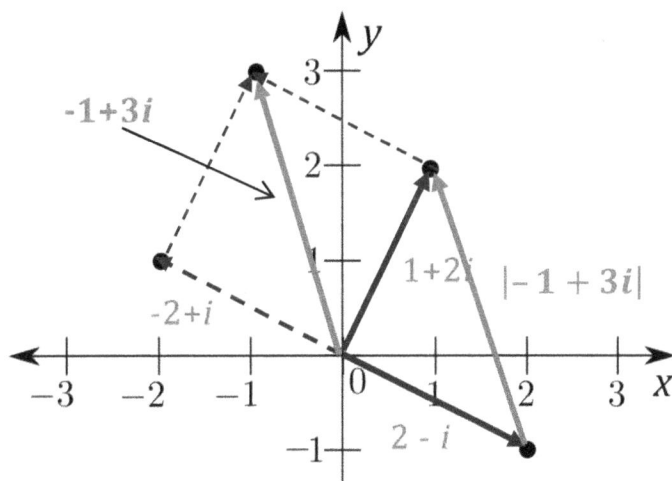

In general, given $z, w \in \mathbb{C}$, we can visualize $z - c$ as a vector that begins at the terminal point of w and ends at the terminal point of z.

Division: If $z \in \mathbb{C}$ and $w \in \mathbb{C}^*$ with $z = a + bi$ and $w = c + di$, then we define the **quotient** $\frac{z}{w}$ by

$$\frac{z}{w} = zw^{-1} = (a + bi)\left(\frac{c}{c^2 + d^2} - \frac{d}{c^2 + d^2}i\right) = \frac{ac + bd}{c^2 + d^2} + \frac{bc - ad}{c^2 + d^2}i.$$

The definition of division in a field unfortunately led to a messy looking formula. However, when actually performing division, there is an easier way to think about it, as we will see below.

The **conjugate** of the complex number $z = a + bi$ is the complex number $\bar{z} = a - bi$.

Exercise 7.8: Find the conjugate of each of the following complex numbers:

1. $1 + i$ _____

2. $2 - 3i$ _____

3. $5i$ _____

4. 7 _____

5. $\sqrt{2} - \sqrt{5}i$ _____

Notes: (1) To take the conjugate of a complex number, we simply negate the imaginary part of the number and leave the real part as it is.

(2) If $z = a + bi \neq 0$, then at least one of a or b is not zero. It follows that $\bar{z} = a - bi$ is also not 0.

(3) The product of a complex number with its conjugate is always a nonnegative real number. Specifically, if $z = a + bi$, then $z\bar{z} = (a + bi)(a - bi) = (a^2 + b^2) + (-ab + ab)i = a^2 + b^2$.

(4) We can change the quotient $\frac{z}{w}$ to standard form by multiplying the numerator and denominator by $\bar{w}$. So, if $z = a + bi$ and $w = c + di$, then we have

$$\frac{z}{w} = \frac{z\bar{w}}{w\bar{w}} = \frac{(a + bi)(c - di)}{(c + di)(c - di)} = \frac{(ac + bd) + (bc - ad)i}{c^2 + d^2} = \frac{ac + bd}{c^2 + d^2} + \frac{bc - ad}{c^2 + d^2}i.$$

Example 7.9: Let $z = 2 - 3i$ and $w = -1 + 5i$. Then

$$\bar{z} = 2 + 3i. \qquad\qquad \bar{w} = -1 - 5i.$$

$$\frac{z}{w} = \frac{z\bar{w}}{w\bar{w}} = \frac{(2 - 3i)(-1 - 5i)}{(-1 + 5i)(-1 - 5i)} = \frac{(-2 - 15) + (-10 + 3)i}{(-1)^2 + 5^2} = \frac{(-17 - 7i)}{1 + 25} = -\frac{17}{26} - \frac{7}{26}i.$$

Exercise 7.10: Compute each of the following quotients:

1. $\frac{1+i}{1-i}$ = _____

2. $\frac{-3-2i}{-5+i}$ = _____

3. $\frac{5i}{9-3i}$ = _____

Absolute Value and Distance

If x and y are real or complex numbers such that $y = x^2$, then we call x a **square root** of y. If x is a positive real number, then we say that x is the **positive square root** of y and we write $x = \sqrt{y}$.

Example 7.11:

1. Since $2^2 = 4$, $2 \in \mathbb{R}$, and $2 > 0$, we see that 2 is the positive square root of 4 and we write $2 = \sqrt{4}$.

2. We have $(-2)^2 = 4$, but $-2 < 0$, and so we **do not** write $-2 = \sqrt{4}$. However, -2 is still a square root of 4, and we can write $-2 = -\sqrt{4}$.

3. Since $i^2 = -1$, we see that i is a square root of -1.

4. Since $(-i)^2 = (-i)(-i) = (-1)(-1)i^2 = 1(-1) = -1$, we see that $-i$ is also a square root of -1.

5. $(1+i)^2 = (1+i)(1+i) = (1-1) + (1+1)i = 0 + 2i = 2i$. So, $1 + i$ is a square root of $2i$.

Exercise 7.12: Find all square roots of i.

_____

_____

_____

_____

The **absolute value** or **modulus** of the complex number $z = a + bi$ is the nonnegative real number

$$|z| = \sqrt{a^2 + b^2} = \sqrt{(\text{Re } z)^2 + (\text{Im } z)^2}.$$

Example 7.13:

1. $|3 + 4i| = \sqrt{3^2 + 4^2} = \sqrt{9 + 16} = \sqrt{25} = \mathbf{5}$.

2. $|-2 + 5i| = \sqrt{(-2)^2 + 5^2} = \sqrt{4 + 25} = \sqrt{\mathbf{29}}$.

3. $|11| = |11 + 0i| = \sqrt{11^2 + 0^2} = \sqrt{11^2} = \mathbf{11}$.

4. $|-3| = |-3 + 0i| = \sqrt{(-3)^2 + 0^2} = \sqrt{9 + 0} = \sqrt{9} = \mathbf{9}$.

5. $|7i| = |0 + 7i| = \sqrt{0^2 + 7^2} = \sqrt{7^2} = \mathbf{7}$.

Exercise 7.14: Compute each of the following

1. $|12 - 5i|$ $\quad = \quad$ _____

2. $|-\sqrt{7}i|$ $\quad = \quad$ _____

3. $|\sqrt{15} - \sqrt{21}i|$ $\quad = \quad$ _____

4. $|0|$ $\quad = \quad$ _____

5. $|\sqrt{6} - \sqrt{11}i|$ $\quad = \quad$ _____

Note: If $z = a + 0i = a$ is a real number, then $|a| = \sqrt{a^2}$. This is equal to a if $a \geq 0$ and $-a$ if $a < 0$.

For example, $|4| = \sqrt{4^2} = \sqrt{16} = 4$ and $|-4| = \sqrt{(-4)^2} = \sqrt{16} = 4 = -(-4)$.

131

The statement "$|a| = -a$ for $a < 0$" often confuses students. This confusion is understandable, as a minus sign is usually used to indicate that an expression is negative, whereas here we are negating a negative number to make it positive. Unfortunately, this is the simplest way to say, "delete the minus sign in front of the number" using basic notation.

Geometrically, the absolute value of a complex number z is the distance between the point z and the origin.

Example 7.15: Which of the following complex numbers is closest to the origin? $1 + 2i$, $-3 + i$, or $-2 + 3i$?

$$|1 + 2i| = \sqrt{1^2 + 2^2} = \sqrt{1 + 4} = \sqrt{5}$$

$$|-3 + i| = \sqrt{(-3)^2 + 1^2} = \sqrt{9 + 1} = \sqrt{10}$$

$$|-2 + 3i| = \sqrt{(-2)^2 + 3^2} = \sqrt{4 + 9} = \sqrt{13}$$

Since $\sqrt{5} < \sqrt{10} < \sqrt{13}$, we see that $1 + 2i$ is closest to the origin.

Notes: (1) Here we have used the following fact: If $a, b \in \mathbb{R}^+$, then $a < b$ if and only if $a^2 < b^2$. Applying this fact to $5 < 10 < 13$, we get $\sqrt{5} < \sqrt{10} < \sqrt{13}$.

(2) Below is a picture displaying $1 + 2i$, $-3 + i$, and $-2 + 3i$ in the Complex Plane.

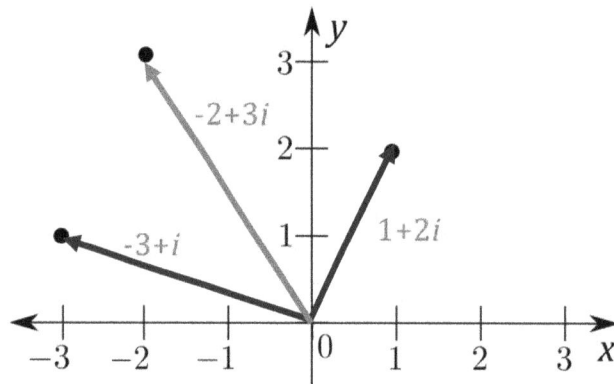

(3) The definition of the absolute value of a complex number is motivated by the Pythagorean Theorem.

As an example, look at $-3 + i$ in the figure below.

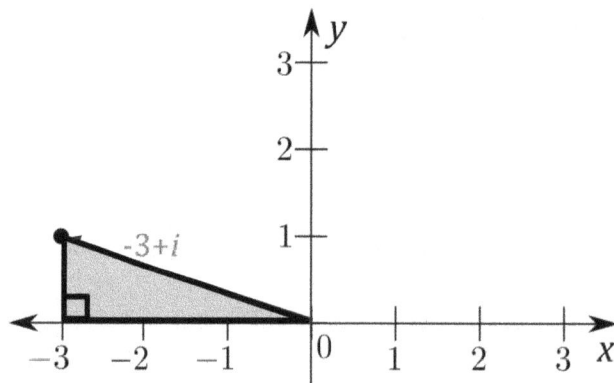

132

Observe that to get from the origin to the point $(-3, 1)$, we move to the left 3 units and then up 1 unit. This gives us a right triangle with legs of lengths 3 and 1. By the Pythagorean Theorem, the hypotenuse has length $\sqrt{3^2 + 1^2} = \sqrt{9 + 1} = \sqrt{10}$.

The **distance** between the complex numbers $z = a + bi$ and $w = c + di$ is

$$d(z, w) = |z - w| = \sqrt{(a - c)^2 + (b - d)^2}.$$

Note: The distance between the real numbers x and y is simply $d(x, y) = |x - y| = \sqrt{(x - y)^2}$.

Example 7.16:

1. The distance between 2 and 5 is $|2 - 5| = |-3| = \mathbf{3}$.

   Observe that the distance between 5 and 2 is the same. Indeed, we have $|5 - 2| = |3| = 3$.

2. The distance between $1 + 2i$ and $2 - i$ is

$$|(1 + 2i) - (2 - i)| = \sqrt{(1 - 2)^2 + \left(2 - (-1)\right)^2} = \sqrt{(-1)^2 + 3^2} = \sqrt{1 + 9} = \sqrt{\mathbf{10}}.$$

Notes: (1) In general, for real numbers x and y, $|x - y| = |y - x|$. This follows immediately from the fact that $y - x = -(x - y)$. The same will be true if x and y are complex numbers.

(2) Take a look one more time at the figure we drew above for $(1 + 2i) - (2 - i) = -1 + 3i$ with the solution vector $z - w = -1 + 3i$ translated so that the directed line segment begins at the terminal point of $w = 2 - i$ and ends at the terminal point of $z = 1 + 2i$. (See the Note after Exercise 7.7)

Notice that the expression for the distance between two complex numbers follows from a simple application of the Pythagorean Theorem. Let's continue to use the same figure to help us see this.

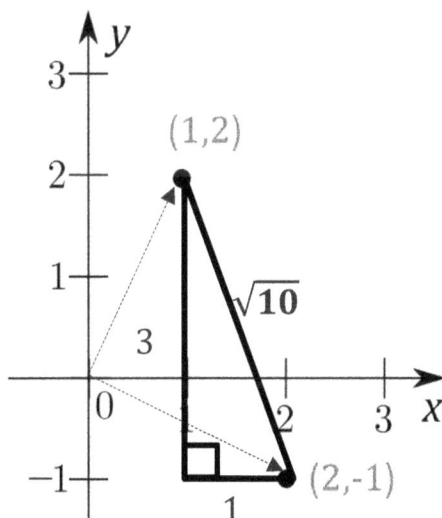

In the figure above, we can get the lengths of the legs of the triangle either by simply counting the units, or by subtracting the appropriate coordinates. For example, the length of the horizontal leg is $2 - 1 = 1$ and the length of the vertical leg is $2 - (-1) = 2 + 1 = 3$. We can then use the Pythagorean Theorem to get the length of the hypotenuse of the triangle: $c = \sqrt{1^2 + 3^2} = \sqrt{1 + 9} = \sqrt{10}$.

Compare this geometric procedure to the formula for distance given above.

Exercise 7.17: For each of the following, compute the distance between the two complex numbers.

1. $z = 1 + i, w = 1 - i$ $=$ _____

2. $z = 2 - 3i, w = 5 + 2i$ $=$ _____

3. $z = 5, w = -i$ $=$ _____

4. $z = 0, w = 8 + 9i$ $=$ _____

Basic Topology of $\mathbb{C}$

Warning: The topology of $\mathbb{C}$ is more complicated than the topology of $\mathbb{R}$. There is no harm in skipping this section and moving on to Lesson 8 if the material is too difficult upon a first reading.

A **circle** in the Complex Plane is the set of all points that are at a fixed distance from a fixed point. The fixed distance is called the **radius** of the circle and the fixed point is called the **center** of the circle.

If a circle has radius $r > 0$ and center $c = a + bi$, then any point $z = x + yi$ on the circle must satisfy $|z - c| = r$, or equivalently, $(x - a)^2 + (y - b)^2 = r^2$.

Note: The equation $|z - c| = r$ says "The distance between z and c is equal to r." In other words, the distance between any point on the circle and the center of the circle is equal to the radius of the circle.

Example 7.18: The circle with equation $|z + 2 - i| = 2$ has center $c = -(2 - i) = -2 + i$ and radius $r = 2$.

Note: $|z + 2 - i| = |z - (-2 + i)|$. So, if we rewrite the equation as $|z - (-2 + i)| = 2$, it is easy to pick out the center and radius of the circle.

A picture of the circle is shown to the right. The center is labeled and a typical radius is drawn.

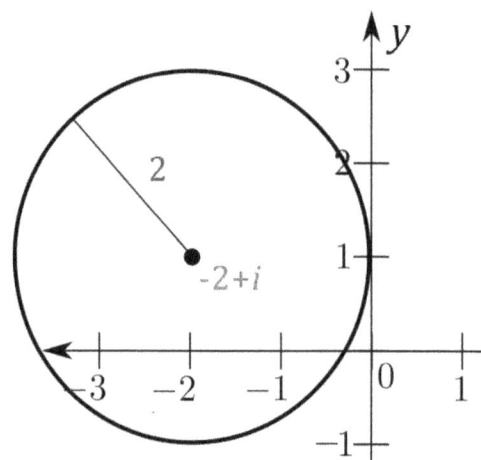

Exercise 7.19: Find the center and radius of the circle with the given equation.

1. $|z - i| = 5$

2. $|z - (2 - 3i)| = 11$

3. $|z + 3 + 2i| = 2$

4. $x^2 + y^2 = 9$

5. $(x - 3)^2 + (y + 1)^2 = 17$

Note: If x and a are real numbers, then the equation $|x - a| = r$ is equivalent to the statement "$x - a = r$ or $x - a = -r$," or equivalently, "$x = a + r$ or $x = a - r$." So, the real-valued version of a "circle" is two points that are equidistant from the "center" of the "circle."

Example 7.20: The equation $|x - 1| = 2$ is equvalent to "$x = 1 + 2 = 3$ or $x = 1 - 2 = -1$". In other words, there are exactly two real numbers that are at a distance of 2 from 1. These two numbers are 3 and -1.

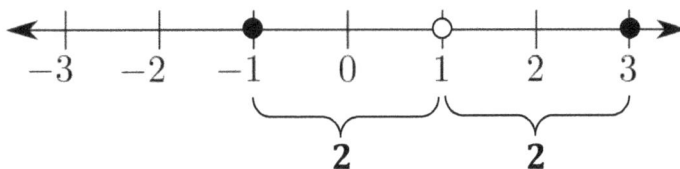

An **open disk** in $\mathbb{C}$ consists of all the points in the interior of a circle. If a is the center of the open disk and r is the radius of the open disk, then any point z inside the disk satisfies $|z - a| < r$.

$N_r(a) = \{z \in \mathbb{C} \mid |z - a| < r\}$ is also called the **r-neighborhood of a**.

Example 7.21: $N_2(-2 + i) = \{z \in \mathbb{C} \mid |z + 2 - i| < 2\}$ is the 2 neighborhood of $-2 + i$ (or equivalently, the open disk with center $-2 + i$ and radius 2). It consists of all points inside the circle $|z + 2 - i| = 2$.

Notes: (1) A picture of the 2-neighborhood of $-2 + i$ is shown to the right. The center is labeled and a typical radius is drawn. We drew the boundary of the disk with dashes to indicate that points on the circle are **not** in the neighborhood and we shaded the interior of the disk to indicate that every point inside the circle is in the neighborhood.

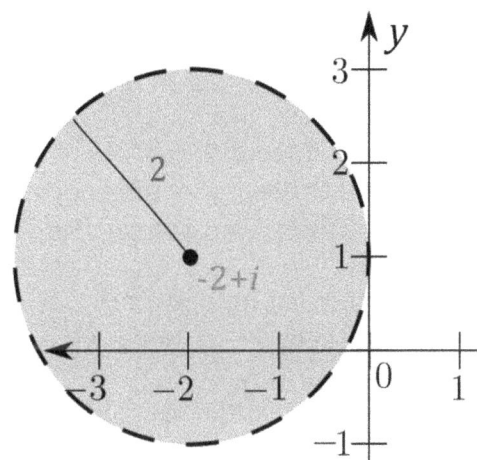

(2) The definitions of open disk and r-neighborhood of a also make sense in $\mathbb{R}$, but the geometry looks a bit different. An open disk in $\mathbb{R}$ is simply an open interval. If x and a are real numbers, then we have

$$x \in N_r(a) \Leftrightarrow |x - a| < r \Leftrightarrow \sqrt{(x - a)^2} < r \Leftrightarrow 0 \leq (x - a)^2 < r^2$$
$$\Leftrightarrow -r < x - a < r \Leftrightarrow a - r < x < a + r \Leftrightarrow x \in (a - r, a + r).$$

So, in $\mathbb{R}$, an r-neighborhood of a is the open interval $N_r(a) = (a - r, a + r)$. Notice that the length (or **diameter**) of this interval is $2r$.

As an example, let's draw a picture of $N_2(1) = (1 - 2, 1 + 2) = (-1, 3)$. Observe that the center of this open disk (or open interval or neighborhood) in $\mathbb{R}$ is the real number 1, the radius of the open disk is 2, and the diameter of the open disk (or length of the interval) is 4.

135

Exercise 7.22: Find the center and radius of each of the following open disks.

1. $N_4(7 - 3i)$
2. $\{z \in \mathbb{C} \mid |z - 1 + i| < 6\}$
3. $\{x + yi \in \mathbb{C} \mid (x + 2)^2 + (y - 1)^2 < 3\}$

A **closed disk** is the interior of a circle together with the circle itself (the **boundary** is included). If a is the center of the closed disk and r is the radius of the closed disk, then any point z inside the closed disk satisfies $|z - a| \le r$.

Exercise 7.23: Use set-builder notation to describe the closed disk with the given center and radius.

1. Center $= 3i$, Radius $= 1$
2. Center $= 8 + i$, Radius $= \sqrt{2}$
3. Center $= -1 + 2i$, Radius $= 1.5$

Notes: (1) When drawing a closed disk, the circle itself would be drawn solid to indicate that all points on the circle are included.

(2) Just like an open disk in $\mathbb{R}$ is an open interval, a closed disk in $\mathbb{R}$ is a closed interval.

A **punctured open disk** consists of all the points in the interior of a circle **except** for the center of the circle. If a is the center of the punctured open disk and r is the radius of the open disk, then any point z inside the punctured disk satisfies $|z - a| < r$ and $z \ne a$.

Note that $z \ne a$ is equivalent to $z - a \ne 0$. In turn, this is equivalent to $|z - a| \ne 0$. Since $|z - a|$ must be nonnegative, $|z - a| \ne 0$ is equivalent to $|z - a| > 0$ or $0 < |z - a|$.

Therefore, a punctured open disk with center a and radius r consists of all points z that satisfy

$$0 < |z - a| < r.$$

$N_r^{\odot}(a) = \{z \mid 0 < |z - a| < r\}$ is also called a **deleted r-neighborhood** of a.

Example 7.24: $N_2^{\odot}(-2 + i) = \{z \in \mathbb{C} \mid 0 < |z + 2 - i| < 2\}$ is the deleted 2 neighborhood of $-2 + i$. It consists of all points inside the circle $|z + 2 - i| = 2$, **except for** $-2 + i$.

Notes: (1) A picture of the deleted 2-neighborhood of $-2 + i$ is shown to the right. Notice that this time we excluded the center of the disk $-2 + i$, as this point is not included in the set.

(2) In $\mathbb{R}$, we have

$$N_r^{\odot}(a) = (a - r, a + r) \setminus \{a\} = (a - r, a) \cup (a, a + r).$$

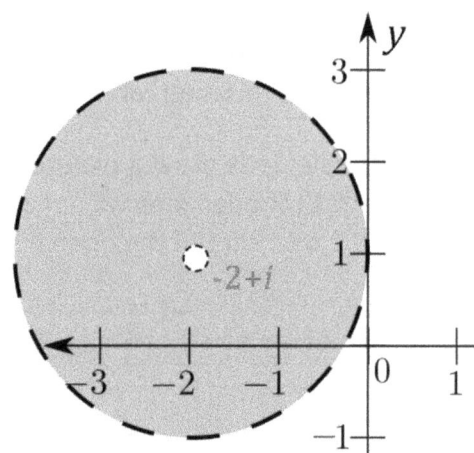

This is the open interval centered at a of length (or diameter) $2r$ with a removed.

Let's draw a picture of $N_2^{\circleddash}(1) = (-1, 3) \setminus \{1\} = (-1, 1) \cup (1, 3)$.

A subset X of $\mathbb{C}$ is said to be **open** if for every complex number $z \in X$, there is an open disk D with $z \in D$ and $D \subseteq X$.

In words, a set is open in $\mathbb{C}$ if every point in the set has "space" all around it inside the set. If you think of each point in the set as an animal, then each animal in the set should be able to move a little in any direction it chooses without leaving the set. Another way to think of this is that no number is right on "the edge" or "the boundary" of the set, about to fall out of it.

Example 7.25:

1. Every open disk D is an open set. To see this, simply observe that if $z \in D$, then D itself is an open disk with $z \in D$ and $D \subseteq D$.

2. A closed disk is **not** an open set because it contains its "boundary." As an example, let's look at the closed unit disk $D = \{z \in \mathbb{C} \mid |z| \leq 1\}$. Let's focus on the point i. First note that $i \in D$ because $|i| = \sqrt{0^2 + 1^2} = \sqrt{1} = 1$ and $1 \leq 1$. However, any open disk N containing i will contain points above i.

3. We can use reasoning similar to that used in 2 to see that if we take any subset of a disk that contains any points on the bounding circle, then that set will **not** be open.

4. $\emptyset$ and $\mathbb{C}$ are both open.

Exercise 7.26: Determine if each of the following sets is open in $\mathbb{C}$.

1. $\{z \in \mathbb{C} \mid |z| < 5\}$
2. $\{z \in \mathbb{C} \mid |z| \leq 7\}$
3. $\{z \in \mathbb{C} \mid |z| > 1\}$
4. $\{z \in \mathbb{C} \mid |z - 1 + i| \leq 3\}$
5. $\{z \in \mathbb{C} \mid |z + 2 - 3i| > 3\}$
6. $\{x + yi \in \mathbb{C} \mid x^2 + (y - 5)^2 < 17\}$
7. $N_4(3 - 5i)$
8. $N_2^{\circleddash}(-2 + i)$

A subset X of $\mathbb{C}$ is said to be **closed** if $\mathbb{C} \setminus X$ is open.

Example 7.27: The closed unit disk $D = \{z \in \mathbb{C} \mid |z| \leq 1\}$ is closed in $\mathbb{C}$ because its complement in $\mathbb{C}$ is $\mathbb{C} \setminus D = \{z \in \mathbb{C} \mid |z| > 1\}$, which is open.

Problem Set 7

Full solutions to these problems are available for free download here:

www.SATPrepGet800.com/PMNR2ZX

LEVEL 1

Let $z = 1 + i$ and $w = 2 - 3i$. Compute each of the following:

1. Re z

2. Im z

3. Re w

4. Im w

5. $z + w$

6. $z - w$

7. $2z$

8. $5w$

9. $2z + 5w$

10. $w - 5z$

11. $\bar{z}$

12. $\bar{w}$

13. $\bar{z} + \bar{w}$

14. $\overline{z + w}$

LEVEL 2

Find all square roots of each of the following complex numbers:

15. 4

16. $2i$

Let $z = -4 - i$ and $w = 3 - 5i$. Compute each of the following:

17. zw

18. $\dfrac{z}{w}$

19. $|z|$

20. $|w|$

21. $|zw|$

22. the distance between z and w.

LEVEL 3

Determine if each of the following sets is open, closed, both, or neither.

23. $\emptyset$

24. $\mathbb{C}$

25. $\{z \in \mathbb{C} \mid |z| > 1\}$

26. $\{z \in \mathbb{C} \mid \operatorname{Im} z \leq -2\}$

27. $\{i^n \mid n \in \mathbb{Z}^+\}$

28. $\{z \in \mathbb{C} \mid 2 < |z - 2| < 4\}$

A point w is an **accumulation point** of a set S of complex numbers if each deleted neighborhood of w contains at least one point in S. Determine the accumulation points of each of the following sets:

29. $\left\{ \dfrac{1}{n} \,\middle|\, n \in \mathbb{Z}^+ \right\}$

30. $\left\{ \dfrac{i}{n} \,\middle|\, n \in \mathbb{Z}^+ \right\}$

31. $\{i^n \mid n \in \mathbb{Z}^+\}$

32. $\left\{ \dfrac{i^n}{n} \,\middle|\, n \in \mathbb{Z}^+ \right\}$

33. $\{z \mid |z| < 1\}$

34. $\{z \mid 0 < |z - 2| \leq 3\}$

35. $\{a + bi \mid a, b \in \mathbb{Z}\}$

36. $\{a + bi \mid a, b \in \mathbb{Q}\}$

LEVEL 4

Let z and w be complex numbers. Verify each of the following:

37. $\operatorname{Re} z = \frac{z + \bar{z}}{2}$

38. $\operatorname{Im} z = \frac{z - \bar{z}}{2i}$

39. $\overline{z + w} = \bar{z} + \bar{w}$

40. $\overline{zw} = \bar{z} \cdot \bar{w}$

41. $\overline{\left(\frac{z}{w}\right)} = \frac{\bar{z}}{\bar{w}}$

42. $z\bar{z} = |z|^2$

43. $|zw| = |z||w|$

44. If $w \neq 0$, then $\left|\frac{z}{w}\right| = \frac{|z|}{|w|}$

45. $\operatorname{Re} z \leq |z|$

46. $\operatorname{Im} z \leq |z|$

47. $|z + w| \leq |z| + |w|$ (This is known as the **Triangle Inequality**.)

48. $\big||z| - |w|\big| \leq |z \pm w| \leq |z| + |w|$.

Let $c = 3 - i$ and $r = \sqrt{5}$. Use set-builder notation to describe each of the following:

49. The circle with center c and radius r.

50. The open disk with center c and radius r.

51. The closed disk with center c and radius r.

52. The punctured open disk with center c and radius r.

LEVEL 5

Verify each of the following:

53. An arbitrary union of open sets in $\mathbb{C}$ is an open set in $\mathbb{C}$.

54. An arbitrary intersection of closed sets in $\mathbb{C}$ is a closed set in $\mathbb{C}$.

A complex number z is an **interior point** of a set S of complex numbers if there is a neighborhood of z that contains only points in S, whereas w is a **boundary point** of S if each neighborhood of w contains at least one point in S and one point not in S. Show the following:

55. A set of complex numbers is open if and only if each point in S is an interior point of S.

56. A set of complex numbers is open if and only if it contains none of its boundary points.

57. A set of complex numbers is closed if and only if it contains all its boundary points.

CHALLENGE PROBLEMS

Recall that a point w is an **accumulation point** of a set S of complex numbers if each deleted neighborhood of w contains at least one point in S.

58. Show that a set of complex numbers is a closed set in $\mathbb{C}$ if and only if it contains all its accumulation points.

59. Show that a set consisting of finitely many complex numbers is a closed set in $\mathbb{C}$.

60. Let $S \subseteq \mathbb{C}$ and let $\overline{S} = \cap\{C \mid S \subseteq C \wedge C \text{ is closed}\}$ ($\overline{S}$ is called the **closure** of S in $\mathbb{C}$). Show that $\overline{S} = S \cup \{z \in \mathbb{C} \mid z \text{ is an accumulation point of } S\}$.

61. Show that $z \in \overline{S}$ if and only if every open disk containing z contains at least one point of S (see Problem 60 above for the definition of $\overline{S}$).

141

LESSON 8
LINEAR ALGEBRA

Matrices

A 2×2 **matrix** over $\mathbb{R}$ is a rectangular array with 2 rows and 2 columns, and **entries** that are real numbers.

Example 8.1: Each of the following is a 2×2 matrix over $\mathbb{R}$.

1. $\begin{bmatrix} 1 & 3 \\ 7 & 5 \end{bmatrix}$

2. $\begin{bmatrix} \sqrt{2} & -3 \\ 5.3 & 0 \end{bmatrix}$

3. $\mathbf{0} = \begin{bmatrix} 0 & 0 \\ 0 & 0 \end{bmatrix}$ (This is known as the 2×2 **zero matrix**.)

4. $I = \begin{bmatrix} 1 & 0 \\ 0 & 1 \end{bmatrix}$ (This is known as the 2×2 **identity matrix**.)

Let $M = \left\{ \begin{bmatrix} a & b \\ c & d \end{bmatrix} \mid a, b, c, d \in \mathbb{R} \right\}$ be the set of all 2×2 matrices over $\mathbb{R}$. We add two such matrices using the rule $\begin{bmatrix} a & b \\ c & d \end{bmatrix} + \begin{bmatrix} e & f \\ g & h \end{bmatrix} = \begin{bmatrix} a+e & b+f \\ c+g & d+h \end{bmatrix}$, and we multiply a matrix by a real number using the rule $k \begin{bmatrix} a & b \\ c & d \end{bmatrix} = \begin{bmatrix} ka & kb \\ kc & kd \end{bmatrix}$ (this type of multiplication is called **scalar multiplication**).

Example 8.2: Let $A = \begin{bmatrix} 1 & 3 \\ 7 & 5 \end{bmatrix}$ and let $B = \begin{bmatrix} 2 & 0 \\ 1 & 1 \end{bmatrix}$. We have the following:

1. $A + B = \begin{bmatrix} 1 & 3 \\ 7 & 5 \end{bmatrix} + \begin{bmatrix} 2 & 0 \\ 1 & 1 \end{bmatrix} = \begin{bmatrix} 1+2 & 3+0 \\ 7+1 & 5+1 \end{bmatrix} = \begin{bmatrix} 3 & 3 \\ 8 & 6 \end{bmatrix}$

2. $A + \mathbf{0} = \begin{bmatrix} 1 & 3 \\ 7 & 5 \end{bmatrix} + \begin{bmatrix} 0 & 0 \\ 0 & 0 \end{bmatrix} = \begin{bmatrix} 1+0 & 3+0 \\ 7+0 & 5+0 \end{bmatrix} = \begin{bmatrix} 1 & 3 \\ 7 & 5 \end{bmatrix}$

3. $2A = 2 \begin{bmatrix} 1 & 3 \\ 7 & 5 \end{bmatrix} = \begin{bmatrix} 2 \cdot 1 & 2 \cdot 3 \\ 2 \cdot 7 & 2 \cdot 5 \end{bmatrix} = \begin{bmatrix} 2 & 6 \\ 14 & 10 \end{bmatrix}$

4. $3B = 3 \begin{bmatrix} 2 & 0 \\ 1 & 1 \end{bmatrix} = \begin{bmatrix} 3 \cdot 2 & 3 \cdot 0 \\ 3 \cdot 1 & 3 \cdot 1 \end{bmatrix} = \begin{bmatrix} 6 & 0 \\ 3 & 3 \end{bmatrix}$

5. $2A + 3B = 2 \begin{bmatrix} 1 & 3 \\ 7 & 5 \end{bmatrix} + 3 \begin{bmatrix} 2 & 0 \\ 1 & 1 \end{bmatrix} = \begin{bmatrix} 2 & 6 \\ 14 & 10 \end{bmatrix} + \begin{bmatrix} 6 & 0 \\ 3 & 3 \end{bmatrix} = \begin{bmatrix} 2+6 & 6+0 \\ 14+3 & 10+3 \end{bmatrix} = \begin{bmatrix} 8 & 6 \\ 17 & 13 \end{bmatrix}$

Exercise 8.3: Let $A = \begin{bmatrix} -2 & 0 \\ -7 & 1 \end{bmatrix}$ and let $B = \begin{bmatrix} 3 & -1 \\ 5 & 5 \end{bmatrix}$. Compute each of the following:

1. $A + B = $ _____

2. $\mathbf{0} + B = $ _____

3. $5A = $ _____

4. $5A + 2B = $ _____

More generally, for $m, n \in \mathbb{Z}^+$, an $m \times n$ **matrix** over a field F is a rectangular array with m rows and n columns, and entries in F. We will say that the **size** of the matrix is $m \times n$ (other authors may refer to $m \times n$ as the **dimensions** of the matrix). For example, the matrix $H = \begin{bmatrix} i & 2 - 5i & \frac{1}{5} \\ -1 & \sqrt{3} & 7 + i \end{bmatrix}$ is a 2×3 matrix over $\mathbb{C}$ (the field of complex numbers). Notice that the first number in the size of the matrix (2 in this example) indicates that the matrix has 2 rows, while the second number in the size of the matrix (3 in this example) indicates that the matrix has 3 columns.

Exercise 8.4: Each of the following is an $m \times n$ matrix over a field F. Determine m, n and F (note that there may be more than one possible answer for F).

1. $A = \begin{bmatrix} 1 & 2 \\ 3 & 4 \end{bmatrix}$ $m =$ ___ $n =$ ___ $F =$ ___

2. $B = \begin{bmatrix} \sqrt{2} & \sqrt{3} & 2 \\ \sqrt{5} & \sqrt{6} & \sqrt{7} \end{bmatrix}$ $m =$ ___ $n =$ ___ $F =$ ___

3. $C = \begin{bmatrix} 1 \\ i \\ -i \\ -1 \end{bmatrix}$ $m =$ ___ $n =$ ___ $F =$ ___

4. $D = \begin{bmatrix} 1 & 2 & 3 & \sqrt{5} \end{bmatrix}$ $m =$ ___ $n =$ ___ $F =$ ___

5. $E = \begin{bmatrix} 1 & 0 & 1 & 0 \\ 2 & -3 & 0 & 0 \\ 0 & 5 & 0 & -2 \end{bmatrix}$ $m =$ ___ $n =$ ___ $F =$ ___

We will generally use a capital letter to represent a matrix (as we have done above), and the corresponding lowercase letter with **double subscripts** to represent the **entries** of the matrix. We use the first subscript for the row and the second subscript for the column. Using the matrix H above as an example, we see that $h_{11} = i$, $h_{12} = 2 - 5i$, $h_{13} = \frac{1}{5}$, $h_{21} = -1$, $h_{22} = \sqrt{3}$, and $h_{23} = 7 + i$.

Exercise 8.5: Let $X = \begin{bmatrix} 1 & 9 & 1 & 8 \\ 7 & -1 & 3 & 4 \\ 2 & 5 & 0 & -6 \end{bmatrix}$. Determine each of the following, if it exists:

1. x_{23} ___

2. x_{32} ___

3. x_{34} ___

4. x_{43} ___

5. x_{33} ___

6. x_{24} ___

7. x_{42} ___

8. What is the size of the matrix X? _____

If A and B are $m \times n$ matrices over a field $\mathbb{F}$, then we get the matrix $A + B$ by adding each entry of A to the corresponding entry of B. Note that we can add two matrices only if they have the same size. For example, if A is a 2×3 matrix and B is a 3×2 matrix, then $A + B$ is **undefined**.

Also, if A is an $m \times n$ matrix and $k \in \mathbb{F}$, then we get the matrix kA by multiplying each entry of A by the scalar k.

Example 8.6:

1. $\begin{bmatrix} 3 & 5 & 1 \\ 2 & 0 & 4 \end{bmatrix} + \begin{bmatrix} 2 & 1 & 5 \\ 7 & 3 & 0 \end{bmatrix} = \begin{bmatrix} 5 & 6 & 6 \\ 9 & 3 & 4 \end{bmatrix}$

2. $\begin{bmatrix} 1 \\ i \\ -i \\ -1 \end{bmatrix} + \begin{bmatrix} i \\ 1 \\ 1 + 2i \\ 2 + i \end{bmatrix} = \begin{bmatrix} 1 + i \\ 1 + i \\ 1 + i \\ 1 + i \end{bmatrix}$

3. $2 \begin{bmatrix} 3 & 5 & 1 \\ 2 & 0 & 4 \end{bmatrix} = \begin{bmatrix} 6 & 10 & 2 \\ 4 & 0 & 8 \end{bmatrix}$

4. $3 \begin{bmatrix} 1 \\ i \\ -i \\ -1 \end{bmatrix} = \begin{bmatrix} 3 \\ 3i \\ -3i \\ -3 \end{bmatrix}$

5. $3 \begin{bmatrix} 1 & 0 & 1 & 0 \\ 2 & -3 & 0 & 0 \\ 0 & 5 & 0 & -2 \end{bmatrix} + 2 \begin{bmatrix} 3 & 7 & 0 & 4 \\ 5 & -1 & 7 & -3 \\ 1 & 2 & 1 & 6 \end{bmatrix}$

$= \begin{bmatrix} 3 & 0 & 3 & 0 \\ 6 & -9 & 0 & 0 \\ 0 & 15 & 0 & -6 \end{bmatrix} + \begin{bmatrix} 6 & 14 & 0 & 8 \\ 10 & -2 & 14 & -6 \\ 2 & 4 & 2 & 12 \end{bmatrix}$

$= \begin{bmatrix} 9 & 14 & 3 & 8 \\ 16 & -11 & 14 & -6 \\ 2 & 19 & 2 & 6 \end{bmatrix}$

6. $\begin{bmatrix} 2 & 1 & 5 \\ 7 & 3 & 0 \end{bmatrix} + \begin{bmatrix} 1 & 2 \\ 3 & 4 \end{bmatrix}$ is **undefined** because the two matrices have different sizes. The first is a 2×3 matrix, whereas the second is a 2×2 matrix.

We will now define the product of an $m \times n$ matrix with an $n \times p$ matrix, where m, n, p are positive integers. Note that if A and B are matrices, then to take the product AB we first insist that the number of columns of A be equal to the number of rows of B (these are the "inner" two numbers in the expressions "$m \times n$" and "$n \times p$").

So, how do we actually multiply two matrices? This is a bit complicated and requires just a little practice. Let's walk through an example while informally describing the procedure, so that we can get a feel for how matrix multiplication works.

Let $A = \begin{bmatrix} 0 & 1 \\ 3 & 2 \end{bmatrix}$ and $B = \begin{bmatrix} 1 & 2 & 0 \\ 0 & 3 & 6 \end{bmatrix}$. Notice that A is a 2×2 matrix and B is a 2×3 matrix. Since A has 2 columns and B has 2 rows, we will be able to multiply the two matrices.

For each row of the first matrix and each column of the second matrix, we add up the products entry by entry. Let's compute the product AB as an example.

$$AB = \begin{bmatrix} 0 & 1 \\ 3 & 2 \end{bmatrix} \cdot \begin{bmatrix} 1 & 2 & 0 \\ 0 & 3 & 6 \end{bmatrix} = \begin{bmatrix} x & y & z \\ u & v & w \end{bmatrix}$$

Since x is in the first row and first column, we use the first row of A and the first column of B to get
$x = \begin{bmatrix} 0 & 1 \end{bmatrix} \begin{bmatrix} 1 \\ 0 \end{bmatrix} = 0 \cdot 1 + 1 \cdot 0 = 0 + 0 = 0$.

Since u is in the second row and first column, we use the second row of A and the first column of B to get $u = \begin{bmatrix} 3 & 2 \end{bmatrix} \begin{bmatrix} 1 \\ 0 \end{bmatrix} = 3 \cdot 1 + 2 \cdot 0 = 3$.

The reader should attempt to follow this procedure to compute the values of the remaining entries. The final product is

$$AB = \begin{bmatrix} 0 & 3 & 6 \\ 3 & 12 & 12 \end{bmatrix}$$

Notes: (1) The product of a **2 × 2** matrix and a 2 × 3 matrix is a 2 × 3 matrix.

(2) More generally, the product of an $m \times n$ matrix and an $n \times p$ matrix is an $m \times p$ matrix. Observe that the inner most numbers (both n) must agree, and the resulting product has dimensions given by the outermost numbers (m and p).

Example 8.7:

1. $\begin{bmatrix} 1 & 2 & 3 & 4 \end{bmatrix} \cdot \begin{bmatrix} 5 \\ 1 \\ -2 \\ 3 \end{bmatrix} = \begin{bmatrix} 1 \cdot 5 + 2 \cdot 1 + 3(-2) + 4 \cdot 3 \end{bmatrix} = \begin{bmatrix} 5 + 2 - 6 + 12 \end{bmatrix} = \begin{bmatrix} 13 \end{bmatrix}$.

   We generally identify a 1×1 matrix with its only entry. So, $\begin{bmatrix} 1 & 2 & 3 & 4 \end{bmatrix} \cdot \begin{bmatrix} 5 \\ 1 \\ -2 \\ 3 \end{bmatrix} = \textbf{13}$.

2. $\begin{bmatrix} 5 \\ 1 \\ -2 \\ 3 \end{bmatrix} \cdot \begin{bmatrix} 1 & 2 & 3 & 4 \end{bmatrix} = \begin{bmatrix} 5 & 10 & 15 & 20 \\ 1 & 2 & 3 & 4 \\ -2 & -4 & -6 & -8 \\ 3 & 6 & 9 & 12 \end{bmatrix}$.

   Notice that $\begin{bmatrix} 1 & 2 & 3 & 4 \end{bmatrix} \cdot \begin{bmatrix} 5 \\ 1 \\ -2 \\ 3 \end{bmatrix} \neq \begin{bmatrix} 5 \\ 1 \\ -2 \\ 3 \end{bmatrix} \cdot \begin{bmatrix} 1 & 2 & 3 & 4 \end{bmatrix}$, and in fact, the two products do not even have the same size. This shows that if AB and BA are both defined, then they **do not** need to be equal.

3. $\begin{bmatrix} 1 & 2 \\ 0 & 1 \end{bmatrix} \cdot \begin{bmatrix} 0 & 2 \\ 3 & 2 \end{bmatrix} = \begin{bmatrix} 0+6 & 2+4 \\ 0+3 & 0+2 \end{bmatrix} = \begin{bmatrix} 6 & 6 \\ 3 & 2 \end{bmatrix}.$

$\begin{bmatrix} 0 & 2 \\ 3 & 2 \end{bmatrix} \cdot \begin{bmatrix} 1 & 2 \\ 0 & 1 \end{bmatrix} = \begin{bmatrix} 0+0 & 0+2 \\ 3+0 & 6+2 \end{bmatrix} = \begin{bmatrix} 0 & 2 \\ 3 & 8 \end{bmatrix}.$

Notice that $\begin{bmatrix} 1 & 2 \\ 0 & 1 \end{bmatrix} \cdot \begin{bmatrix} 0 & 2 \\ 3 & 2 \end{bmatrix} \ne \begin{bmatrix} 0 & 2 \\ 3 & 2 \end{bmatrix} \cdot \begin{bmatrix} 1 & 2 \\ 0 & 1 \end{bmatrix}.$

This shows that even if A and B are square matrices of the same size, in general $AB \ne BA$. So, matrix multiplication is **not** commutative.

Exercise 8.8: Compute the following matrix products:

1. $[1 \ \ 3 \ \ 4 \ \ 0] \cdot \begin{bmatrix} 2 \\ 1 \\ 3 \\ 9 \end{bmatrix} =$ _____

2. $\begin{bmatrix} 2 \\ 1 \\ 3 \\ 9 \end{bmatrix} \cdot [1 \ \ 3 \ \ 4 \ \ 0] =$ _____

3. $\begin{bmatrix} 1 & 2 \\ 3 & 4 \end{bmatrix} \cdot \begin{bmatrix} 5 & 6 \\ 0 & 1 \end{bmatrix} =$ _____

4. $\begin{bmatrix} 5 & 6 \\ 0 & 1 \end{bmatrix} \cdot \begin{bmatrix} 1 & 2 \\ 3 & 4 \end{bmatrix} =$ _____

Vector Spaces Over Fields

A **vector space** over a field $\mathbb{F}$ is a set V together with a binary operation $+$ on V (called **addition**) and an operation called **scalar multiplication** satisfying:

(1) $(V, +)$ is a commutative group.

(2) **(Closure under scalar multiplication)** For all $k \in \mathbb{F}$ and $v \in V$, $kv \in V$.

(3) **(Scalar multiplication identity)** If 1 is the multiplicative identity of $\mathbb{F}$ and $v \in V$, then $1v = v$.

(4) **(Associativity of scalar multiplication)** For all $j, k \in \mathbb{F}$ and $v \in V$, $(jk)v = j(kv)$.

(5) **(Distributivity of 1 scalar over 2 vectors)** For all $k \in \mathbb{F}$ and $v, w \in V$, $k(v + w) = kv + kw$.

(6) **(Distributivity of 2 scalars over 1 vector)** For all $j, k \in \mathbb{F}$ and $v \in V$, $(j + k)v = jv + kv$.

Notes: (1) Recall from Lesson 3 that $(V, +)$ a commutative group means the following:

- **(Closure)** For all $v, w \in V$, $v + w \in V$.

- **(Associativity)** For all $v, w, u \in V$, $(v + w) + u = v + (w + u)$.

- **(Commutativity)** For all $v, w \in V$, $v + w = w + v$.

- **(Identity)** There exists an element $0 \in V$ such that for all $v \in V$, $0 + v = v + 0 = v$.

- **(Inverse)** For each $v \in V$, there is $-v \in V$ such that $v + (-v) = (-v) + v = 0$.

(2) The fields that we are familiar with are $\mathbb{Q}$ (the field of rational numbers), $\mathbb{R}$ (the field of real numbers), and $\mathbb{C}$ (the field of complex numbers). For our purposes here, we can always assume that $\mathbb{F}$ is one of these three fields.

(3) Note that a vector space consists of (i) a set of **vectors**, (ii) a field, and (iii) two operations called **addition** and **scalar multiplication**.

Example 8.9: Consider the set $\mathbb{C}$ of complex numbers together with the usual definition of addition. That is, if $z = a + bi$ and $w = c + di$, then $z + w = (a + c) + (b + d)i$. Let's also consider another operation, which we will call **scalar multiplication**. For each $k \in \mathbb{R}$ and $z = a + bi \in \mathbb{C}$, we define kz to be $ka + kbi$.

The operation of scalar multiplication is a little different from other types of operations we have looked at previously because instead of multiplying two elements from $\mathbb{C}$ together, we are multiplying an element of $\mathbb{R}$ with an element of $\mathbb{C}$. In this case, we will call the elements of $\mathbb{R}$ **scalars**.

Let's now check that with these two operations, $\mathbb{C}$ is a vector space over $\mathbb{R}$.

1. $(\mathbb{C}, +)$ **is a commutative group.** In other words, for addition in $\mathbb{C}$, we have closure, associativity, commutativity, an identity element (called 0), and the inverse property (the inverse of $a + bi$ is $-a - bi$). This follows immediately from the fact that $(\mathbb{C}, +, \cdot)$ is a field. When we choose to think of $\mathbb{C}$ as a vector space, we will "forget about" the multiplication in $\mathbb{C}$, and just consider $\mathbb{C}$ together with addition. In doing so, we lose much of the field structure of the complex numbers, but we retain the group structure of $(\mathbb{C}, +)$.

2. $\mathbb{C}$ **is closed under scalar multiplication.** That is, **for all $k \in \mathbb{R}$ and $z \in \mathbb{C}$, we have $kz \in \mathbb{C}$.** To see this, let $z = a + bi \in \mathbb{C}$ and let $k \in \mathbb{R}$. Then, by definition, $kz = ka + kbi$. Since $a, b \in \mathbb{R}$, and $\mathbb{R}$ is closed under multiplication, $ka \in \mathbb{R}$ and $kb \in \mathbb{R}$. It follows that $ka + kbi \in \mathbb{C}$.

3. $1z = z$. To see this, consider $1 \in \mathbb{R}$ and let $z = a + bi \in \mathbb{C}$. Then, since 1 is the multiplicative identity for $\mathbb{R}$, we have $1z = 1a + 1bi = a + bi = z$.

4. **For all $j, k \in \mathbb{R}$ and $z \in \mathbb{C}$, $(jk)z = j(kz)$ (Associativity of scalar multiplication).** To see this, let $j, k \in \mathbb{R}$ and $z = a + bi \in \mathbb{C}$. Then since multiplication is associative in $\mathbb{R}$, we have

$$(jk)z = (jk)(a + bi) = (jk)a + (jk)bi = j(ka) + j(kb)i = j(ka + kbi) = j(kz).$$

5. **For all $k \in \mathbb{R}$ and $z, w \in \mathbb{C}$, $k(z + w) = kz + kw$ (Distributivity of 1 scalar over 2 vectors).** To see this, let $k \in \mathbb{R}$ and $z = a + bi, w = c + di \in \mathbb{C}$. Then since multiplication distributes over addition in $\mathbb{R}$, we have

$$k(z + w) = k\big((a + bi) + (c + di)\big) = k\big((a + c) + (b + d)i\big) = k(a + c) + k(b + d)i$$
$$= (ka + kc) + (kb + kd)i = (ka + kbi) + (kc + kdi) = k(a + bi) + k(c + di) = kz + kw.$$

6. **For all $j, k \in \mathbb{R}$ and $z \in \mathbb{C}$, $(j + k)z = jz + kz$ (Distributivity of 2 scalars over 1 vector).** To see this, let $j, k \in \mathbb{R}$ and $z = a + bi \in \mathbb{C}$. Then since multiplication distributes over addition in $\mathbb{R}$, we have

$$(j + k)z = (j + k)(a + bi) = (j + k)a + (j + k)bi = (ja + ka) + (jb + kb)i$$
$$= (ja + jbi) + (ka + kbi) = j(a + bi) + k(a + bi) = jz + kz.$$

Note: We started with the example of $\mathbb{C}$ as a vector space over $\mathbb{R}$ because it has a geometric interpretation where we can draw simple pictures to visualize what the vector space looks like. Recall from Lesson 7 that we can think of the complex number $a + bi$ as a directed line segment (or **vector**) in the complex plane that begins at the origin and terminates at the point (a, b).

For example, pictured to the right, we can see the vectors $i = 0 + 1i$, $1 + 2i$, and $2 = 2 + 0i$ in the complex plane.

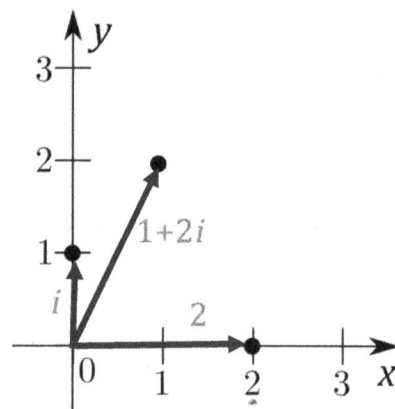

Recall that we visualize the sum of two vectors as the vector starting at the origin that is the diagonal of the parallelogram formed from the original vectors. We see this in the first figure on the left below. In this figure, we have removed the complex plane and focused on the vectors $1 + 2i$ and 2, together with their sum $(1 + 2i) + (2 + 0i) = (1 + 2) + (2 + 0)i = 3 + 2i$.

A second way to visualize the sum of two vectors is to translate one of the vectors so that its initial point coincides with the terminal point of the other vector. The sum of the two vectors is then the vector whose initial point coincides with the initial point of the "unmoved" vector and whose terminal point coincides with the terminal point of the "moved" vector. We see two ways to do this in the center and rightmost figures below.

Technically speaking, the center figure shows the sum $(1 + 2i) + 2$ and the rightmost figure shows the sum $2 + (1 + 2i)$. If we superimpose one figure on top of the other, we can see strong evidence that commutativity holds for addition.

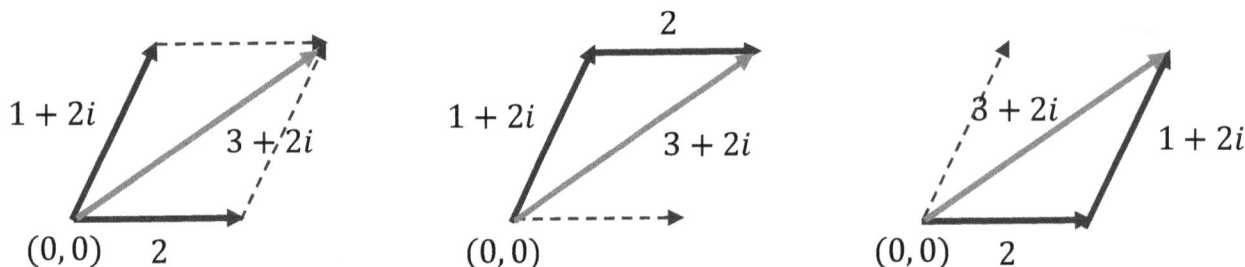

We can visualize a scalar multiple of a vector as follows: (i) if k is a positive real number and $z \in \mathbb{C}$, then the vector kz points in the same direction as z and has a length that is k times the length of z; (ii) if k is a negative real number and $z \in \mathbb{C}$, then the vector kz points in the direction opposite of z and has a length that is $|k|$ times the length of z; (iii) if $k = 0$ and $z \in \mathbb{C}$, then kz is a point.

In the figures below, we have a vector $z \in \mathbb{C}$, together with several scalar multiples of z.

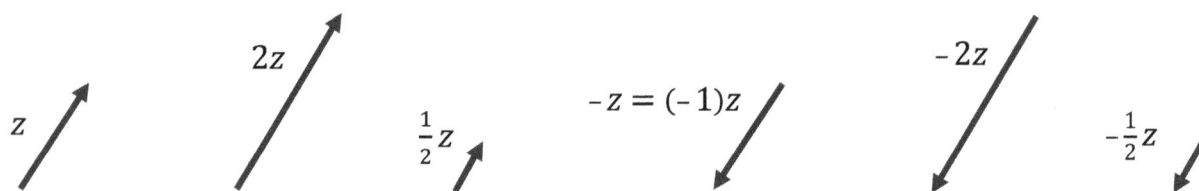

Exercise 8.10: Consider the set M of all 2×2 matrices over $\mathbb{R}$ with addition and scalar multiplication defined as usual. In other words, $\begin{bmatrix} a & b \\ c & d \end{bmatrix} + \begin{bmatrix} e & f \\ g & h \end{bmatrix} = \begin{bmatrix} a+e & b+f \\ c+g & d+h \end{bmatrix}$ and $k\begin{bmatrix} a & b \\ c & d \end{bmatrix} = \begin{bmatrix} ka & kb \\ kc & kd \end{bmatrix}$. Verify each of the following:

1. M is closed under addition.

2. Addition is associative in M.

3. Addition is commutative in M.

4. There is an identity $\mathbf{0}$ in M.

5. Each matrix A in M has an additive inverse in M.

6. $(M, +)$ is a commutative group. _____

7. M is closed under scalar multiplication.

_____

8. For each matrix A in M, $1 \cdot A = A$.

_____

9. For all real numbers j and k, and each matrix A in M, $(jk)A = j(kA)$.

_____

10. For all real numbers k and all matrices A and B in M, $k(A + B) = kA + kB$.

_____

11. For all real numbers j and k, and each matrix A in M, $(j + k)A = jA + kA$.

_____

12. M is a vector space over $\mathbb{R}$. _____

Let's finish this lesson with a few more examples of vector spaces.

Example 8.11:

1. Let $\mathbb{R}^2$ be the set of all ordered pairs of real numbers. That is, $\mathbb{R}^2 = \{(a, b) \mid a, b \in \mathbb{R}\}$ We define **addition** by $(a, b) + (c, d) = (a + c, b + d)$. We define **scalar multiplication** by $k(a, b) = (ka, kb)$ for each $k \in \mathbb{R}$. With these definitions, $\mathbb{R}^2$ is a vector space over $\mathbb{R}$.

   Notice that $\mathbb{R}^2$ looks just like $\mathbb{C}$. In fact, (a, b) is sometimes used as another notation for $a + bi$.

Therefore, the verification that $\mathbb{R}^2$ is a vector space over $\mathbb{R}$ is nearly identical to what we did in Example 8.8 above.

We can visualize elements of $\mathbb{R}^2$ as points or vectors in a plane in exactly the same way that we visualize complex numbers as points or vectors in the Complex Plane.

2. $\mathbb{R}^3 = \{(a, b, c) \mid a, b, c \in \mathbb{R}\}$ is a vector space over $\mathbb{R}$, where we define addition and scalar multiplication by $(a, b, c) + (d, e, f) = (a + d, b + e, c + f)$ and $k(a, b, c) = (ka, kb, kc)$, respectively.

   We can visualize elements of $\mathbb{R}^3$ as points in space in a way similar to visualizing elements of $\mathbb{R}^2$ and $\mathbb{C}$ as points in a plane.

3. More generally, we can let $\mathbb{R}^n = \{(a_1, a_2, \ldots, a_n) \mid a_i \in \mathbb{R} \text{ for each } i = 1, 2, \ldots, n\}$. Then $\mathbb{R}^n$ is a vector space over $\mathbb{R}$, where we define addition and scalar multiplication by

$$(a_1, a_2, \ldots, a_n) + (b_1, b_2, \ldots, b_n) = (a_1 + b_1, a_2 + b_2, \ldots, a_n + b_n).$$
$$k(a_1, a_2, \ldots, a_n) = (ka_1, ka_2, \ldots, ka_n).$$

4. More generally still, if $\mathbb{F}$ is any field (for our purposes, we can think of $\mathbb{F}$ as $\mathbb{Q}$, $\mathbb{R}$, or $\mathbb{C}$), we let $\mathbb{F}^n = \{(a_1, a_2, \ldots, a_n) \mid a_i \in \mathbb{F} \text{ for each } i = 1, 2, \ldots, n\}$. Then $\mathbb{F}^n$ is a vector space over $\mathbb{F}$, where we define addition and scalar multiplication by

$$(a_1, a_2, \ldots, a_n) + (b_1, b_2, \ldots, b_n) = (a_1 + b_1, a_2 + b_2, \ldots, a_n + b_n).$$
$$k(a_1, a_2, \ldots, a_n) = (ka_1, ka_2, \ldots, ka_n).$$

Notes: (1) Ordered pairs have the property that $(a, b) = (c, d)$ if and only if $a = c$ and $b = d$. So, for example, $(1,2) \neq (2,1)$. Compare this to the unordered pair (or set) $\{1, 2\}$. Recall that a set is determined by its elements and not the order in which the elements are listed. So, $\{1, 2\} = \{2, 1\}$.

(2) $(a_1, a_2, \ldots, a_n)$ is called an **n-tuple**. So, $\mathbb{R}^n$ consists of all n-tuples of elements from $\mathbb{R}$, and more generally, $\mathbb{F}^n$ consists of all n-tuples of elements from the field $\mathbb{F}$.

For example, $\left(3, 2 - i, \sqrt{2} + \sqrt{3}i, -3i\right) \in \mathbb{C}^4$ and $\left(1, \frac{1}{2}, \frac{1}{3}, \frac{1}{4}, \frac{1}{5}, \frac{1}{6}, \frac{1}{7}, \frac{1}{8}\right) \in \mathbb{Q}^8$ (and since $\mathbb{Q}^8 \subseteq \mathbb{R}^8 \subseteq \mathbb{C}^8$, we can also say that this 8-tuple is in $\mathbb{R}^8$ or $\mathbb{C}^8$).

(3) Similar to what we said in Note 1, we have $(a_1, a_2, \ldots, a_n) = (b_1, b_2, \ldots, b_n)$ if and only if $a_i = b_i$ for all $i = 1, 2, \ldots, n$. So, for example, $\left(2, 5, \sqrt{2}, \sqrt{2}\right)$ and $\left(2, \sqrt{2}, 5, \sqrt{2}\right)$ are distinct elements from $\mathbb{R}^4$.

(4) You will be asked to verify that $\mathbb{F}^n$ is a vector space over the field $\mathbb{F}$ in Problem 45 through 56 below.

LEVEL 1

Let $A = \begin{bmatrix} 1 & 6 & 3 & 4 \\ 2 & -1 & 8 & -3 \\ 5 & -6 & 7 & 20 \end{bmatrix}$. Determine each of the following, if it exists.

1. a_{14}

2. a_{42}

3. a_{31}

4. $a_{11} + a_{22} + a_{33}$

5. $a_{12} - a_{21}$

6. $2A$

7. $-3A$

8. What is the size of the matrix A?

Let $B = \begin{bmatrix} 0 & 2 & 0 \\ 1 & 15 & 1 \\ 8 & 7 & 5 \\ 6 & 9 & 3 \end{bmatrix}$. Determine each of the following, if it exists.

9. b_{14}

10. b_{42}

11. b_{31}

12. $b_{11} + b_{22} + b_{33}$

13. $b_{12} - b_{21}$

14. $4B$

15. $-2B$

16. What is the size of the matrix B?

Compute each of the following, if possible:

17. $\begin{bmatrix} 4 & 3 \\ 2 & 1 \end{bmatrix} + \begin{bmatrix} 1 & 2 \\ 3 & 4 \end{bmatrix}$

18. $\begin{bmatrix} 1 & 0 & -1 \\ 0 & 1 & 1 \end{bmatrix} + \begin{bmatrix} 1 & 0 & 2 \\ 0 & 1 & 3 \end{bmatrix}$

19. $\begin{bmatrix} 1 & 0 & -1 \\ 0 & 1 & 1 \end{bmatrix} + \begin{bmatrix} 1 & 2 \\ 3 & 4 \end{bmatrix}$

20. $-2\begin{bmatrix} 4 & 3 \\ 2 & 1 \end{bmatrix} + 5\begin{bmatrix} 1 & 2 \\ 3 & 4 \end{bmatrix}$

21. $\begin{bmatrix} 1 & 2 & 3 & 4 \\ 1 & -2 & 3 & -4 \\ 1 & -2 & 3 & 4 \end{bmatrix} + \begin{bmatrix} 2 & 3 & 5 & -3 \\ 2 & -1 & 8 & -1 \\ 5 & -2 & 7 & 3 \end{bmatrix}$

22. $3\begin{bmatrix} 1 & 2 & 3 & 4 \\ 1 & -2 & 3 & -4 \\ 1 & -2 & 3 & 4 \end{bmatrix} + 4\begin{bmatrix} 2 & 3 & 5 & -3 \\ 2 & -1 & 8 & -1 \\ 5 & -2 & 7 & 3 \end{bmatrix}$

23. $\begin{bmatrix} 1 \\ i \\ -i \\ -1 \end{bmatrix} + \begin{bmatrix} 1 & i & i & 1 \end{bmatrix}$

24. $\begin{bmatrix} i & i & i & i \end{bmatrix} + \begin{bmatrix} 1+2i & 1+3i & 1+4i & 1+5i \end{bmatrix}$

25. $5\begin{bmatrix} i & i & i & i \end{bmatrix} + 7\begin{bmatrix} 1+2i & 1+3i & 1+4i & 1+5i \end{bmatrix}$

Compute each of the following, if possible:

26. $\begin{bmatrix} 2 & 0 & -3 \\ 0 & 1 & 4 \end{bmatrix} \cdot \begin{bmatrix} 1 & 1 & 3 & 0 \\ 1 & -4 & 2 & 0 \\ 2 & 0 & 1 & -4 \end{bmatrix}$

27. $\begin{bmatrix} 3 & -1 & 5 \end{bmatrix} \cdot \begin{bmatrix} -4 \\ -7 \\ 2 \end{bmatrix}$

28. $\begin{bmatrix} -4 \\ -7 \\ 2 \end{bmatrix} \cdot \begin{bmatrix} 3 & -1 & 5 \end{bmatrix}$

29. $\begin{bmatrix} a & b & c \\ d & e & f \\ g & h & i \end{bmatrix} \cdot \begin{bmatrix} 1 & 0 & 1 \\ 0 & 2 & 0 \\ 3 & 1 & 4 \end{bmatrix}$

30. $\begin{bmatrix} 1 & 0 & -1 \\ 0 & 1 & 1 \end{bmatrix} \cdot \begin{bmatrix} 1 & 0 & 2 \\ 0 & 1 & 3 \end{bmatrix}$

Questions 31 through 44 require the following definition:

Let V be a vector space over a field $\mathbb{F}$ and let $U \subseteq V$. We say that U is a **subspace** of V if (i) $0 \in U$, (ii) for all $v, w \in U$, $v + w \in U$, and (iii) for all $v \in U$ and $k \in \mathbb{F}$, $kv \in U$.

Determine if each of the following subsets of $\mathbb{R}^2$ is a subspace of $\mathbb{R}^2$ (you may use Problem 59 below):

31. $A = \{(x, y) \mid x + y = 0\}$

32. $B = \{(x, y) \mid xy = 0\}$

33. $C = \{(x, y) \mid 2x = 3y\}$

34. $D = \{(x, y) \mid x \in \mathbb{Q}\}$

35. $E = \{(a, 0) \mid a \in \mathbb{R}\}$

Let U be a subspace of a vector space V over the field $\mathbb{F}$. Explain why each of the following is true

36. Addition is associative in U.

37. Addition is commutative in U.

38. Each element in U has an additive inverse in U (you may use Problem 60 below).

39. $(U, +)$ is a commutative group.

40. For each $x \in U$, $1x = x$.

41. For all $j, k \in \mathbb{F}$ and each $x \in U$, $(jk)x = j(kx)$.

42. For each $k \in \mathbb{F}$ and all $x, y \in U$, $k(x + y) = kx + ky$.

43. For all $j, k \in \mathbb{F}$ and each matrix $x \in U$, $(j + k)x = jx + kx$.

44. U is a vector space over $\mathbb{F}$.

LEVEL 4

Let $\mathbb{F}$ be a field and let $\mathbb{F}^n = \{(a_1, a_2, \ldots, a_n) \mid a_i \in \mathbb{F} \text{ for each } i = 1, 2, \ldots, n\}$. Define addition and scalar multiplication on $\mathbb{F}^n$ as follows:

$$(a_1, a_2, \ldots, a_n) + (b_1, b_2, \ldots, b_n) = (a_1 + b_1, a_2 + b_2, \ldots, a_n + b_n).$$
$$k(a_1, a_2, \ldots, a_n) = (ka_1, ka_2, \ldots, ka_n).$$

Explain why each of the following is true

45. $\mathbb{F}^n$ is closed under addition.

46. Addition is associative in $\mathbb{F}^n$.

47. Addition is commutative in $\mathbb{F}^n$.

48. There is an identity $\mathbf{0}$ in $\mathbb{F}^n$.

49. Each element in $\mathbb{F}^n$ has an additive inverse in $\mathbb{F}^n$.

50. $(\mathbb{F}^n, +)$ is a commutative group.

51. $\mathbb{F}^n$ is closed under scalar multiplication.

52. For each $x \in \mathbb{F}^n$, $1x = x$.

53. For all $j, k \in \mathbb{F}$ and each $x \in \mathbb{F}^n$, $(jk)x = j(kx)$.

54. For each $k \in \mathbb{F}$ and all $x, y \in \mathbb{F}^n$, $k(x + y) = kx + ky$.

55. For all $j, k \in \mathbb{F}$ and each matrix $x \in \mathbb{F}^n$, $(j + k)x = jx + kx$.

56. $\mathbb{F}^n$ is a vector space over $\mathbb{F}$.

LEVEL 5

Let V be a vector space over a field $\mathbb{F}$. Explain why each of the following is true:

57. For every $v \in V$, $-(-v) = v$.

58. For every $v \in V$, $0v = 0$.

59. For every $k \in \mathbb{F}$, $k \cdot 0 = 0$.

60. For every $v \in V$, $-1v = -v$.

Let $A = \begin{bmatrix} a & b \\ c & d \end{bmatrix}$ be a 2×2 matrix. We say that A is **invertible** if there is a 2×2 matrix B such that $AB = I$ and $BA = I$, where $I = \begin{bmatrix} 1 & 0 \\ 0 & 1 \end{bmatrix}$. Verify each of the following:

61. The inverse of $A = \begin{bmatrix} 1 & 1 \\ 0 & 1 \end{bmatrix}$ is $B = \begin{bmatrix} 1 & -1 \\ 0 & 1 \end{bmatrix}$.

62. The zero matrix $\mathbf{0} = \begin{bmatrix} 0 & 0 \\ 0 & 0 \end{bmatrix}$ is **not** invertible.

63. If $ad - bc \neq 0$, then $A = \begin{bmatrix} a & b \\ c & d \end{bmatrix}$ is invertible. In this case, what is the multiplicative inverse of A?

CHALLENGE PROBLEMS

64. If $ad - bc = 0$, then $A = \begin{bmatrix} a & b \\ c & d \end{bmatrix}$ is **not** invertible.

65. Let $n \in \mathbb{Z}^+$ and let M_n be the set of $n \times n$ matrices over a field $\mathbb{F}$. Show that $(M_n, +, \cdot)$ is a ring, where $+$ and $\cdot$ are matrix addition and matrix multiplication, respectively.

Recall that a subset U of a vector space V over a field $\mathbb{F}$ is a **subspace** of V if (i) $0 \in U$, (ii) for all $v, w \in U$, $v + w \in U$, and (iii) for all $v \in U$ and $k \in \mathbb{F}$, $kv \in U$.

66. Show that an arbitrary intersection of subspaces of a vector space V over a field $\mathbb{F}$ is a subspace of V.

67. Let U and W be subspaces of a vector space V. Show that $U \cup W$ is **not** necessarily a subspace of V. Find conditions on U and W that will guarantee that $U \cup W$ will be a subspace of V.

68. Find an example of vector spaces U and V with $U \subseteq V$ such that U is closed under scalar multiplication, but U is not a subspace of V.

Lesson 1

Exercise 1.3:

1. This is a statement. Either the check is in the mail (in which case the statement is true) or the check is not in the mail (in which case the statement is false).

2. This is **not** a statement. It is a **question**.

3. This is a statement. This statement happens to be false.

4. This is **not** a statement. It is a **command**. This particular command is an **idiom**. This means that the actual meaning of the sentence cannot be deduced from the individual words in the sentence. In this particular case, the meaning of the sentence is that you shouldn't accept something as determined before it has actually occurred. Notice how the actual meaning of the sentence does not involve chickens or hatching.

5. This is a statement. Either Odin is chasing a mouse or he is not. **Side note:** in case you haven't figured it out from the context of the statement, Odin is a cat.

Exercise 1.6:

1. This is an **atomic statement**.

2. This is a **compound statement**. It uses the logical connective "or."

3. This is a **compound statement**. It uses the logical connective "not."

4. This is an **atomic statement**. Even though the word "and" appears in the statement, here it is part of the name of the show. It is not being used as a logical connective.

5. This is a **compound statement**. It uses the logical connective "if...then."

6. This is an **atomic statement**.

7. This is a **compound** statement. It actually uses two connectives: "or" and "not."

8. This is a **compound statement**. It uses the logical connective "if and only if."

9. This is an **atomic statement**. Even though the words "and" and "or" appear in the statement, they are **not** being used as logical connectives.

10. This is a **compound statement**. Like part 7 above, it uses two connectives: "and" and "not." Note that in sentential logic the word "but" has the same meaning as the word "and." In English, the word "but" is used to introduce contrast with the part of the sentence that has already been mentioned. However, logically it is no different from "and."

Exercise 1.10: There are **eight** possible truth assignments for this list of propositional variables. We can visualize this list of truth assignments with the following table:

p	q	r
T	T	T
T	T	F
T	F	T
T	F	F
F	T	T
F	T	F
F	F	T
F	F	F

Exercise 1.12:

1. $p \wedge q \equiv T \wedge T \equiv \mathbf{T}$.

2. $p \wedge q \equiv F \wedge F \equiv \mathbf{F}$.

3. $p \wedge q \equiv F \wedge T \equiv \mathbf{F}$.

p	q	$p \wedge q$
T	T	T
T	F	F
F	T	F
F	F	F

Exercise 1.13:

1. $p \vee q \equiv T \vee T \equiv \mathbf{T}$.

2. $p \vee q \equiv F \vee F \equiv \mathbf{F}$.

3. $p \vee q \equiv T \vee F \equiv \mathbf{T}$.

4. $p \vee q \equiv F \vee T \equiv \mathbf{T}$.

p	q	$p \vee q$
T	T	T
T	F	T
F	T	T
F	F	F

Exercise 1.14:

1. $p \rightarrow q \equiv T \rightarrow T \equiv \mathbf{T}$.

2. $p \rightarrow q \equiv F \rightarrow F \equiv \mathbf{T}$.

3. $p \rightarrow q \equiv T \rightarrow F \equiv \mathbf{F}$.

4. $p \rightarrow q \equiv F \rightarrow T \equiv \mathbf{T}$.

p	q	$p \rightarrow q$
T	T	T
T	F	F
F	T	T
F	F	T

Exercise 1.15:

1. $p \leftrightarrow q \equiv T \leftrightarrow T \equiv \mathbf{T}$.

2. $p \leftrightarrow q \equiv F \leftrightarrow F \equiv \mathbf{T}$.

3. $p \leftrightarrow q \equiv T \leftrightarrow F \equiv \mathbf{F}$.

4. $p \leftrightarrow q \equiv F \leftrightarrow T \equiv \mathbf{F}$.

p	q	$p \leftrightarrow q$
T	T	T
T	F	F
F	T	F
F	F	T

Exercise 1.17: $\neg p \equiv \neg F \equiv T$. ◄─────

p	$\neg p$
T	F
F	T

Exercise 1.19: Note that p and q are both false.

1. $p \to q$ represents **"If frogs are birds, then $2 < 1$."** Since p is false, $p \to q$ is **true**.

2. $\neg p \lor q$ represents the statement **"Frogs are not birds or $2 < 1$."** Since $\neg p$ is true, $\neg p \lor q$ is **true**. Note once again that $\neg p \lor q$ always means $(\neg p) \lor q$. In general, without parentheses present, we always apply negation before any of the other connectives.

3. $p \leftrightarrow q$ represents **"Frogs are birds if and only if $2 < 1$."** Since p and q are both false, $p \leftrightarrow q$ is **true**.

4. $(p \to q) \land (q \to p)$ represents **"If frogs are birds, then $2 < 1$ and if $2 < 1$, then frogs are birds."** Since p is false, $p \to q$ is true. Since q is false, $q \to p$ is true. Since $p \to q$ and $q \to p$ are both true, $(p \to q) \land (q \to p)$ is **true**.

5. $\neg(p \land q)$ represents the statement **"It is not the case that both frogs are birds and $2 < 1$."** Since p and q are both false, $p \land q$ is false. It follows that $\neg(p \land q)$ is **true**.

6. $\neg p \lor \neg q$ represents the statement **"Frogs are not birds or 2 is not less than 1."** Since p and q are both false, $\neg p$ and $\neg q$ are both true. It follows that $\neg p \lor \neg q$ is **true**.

Exercise 1.21:

1.

p	q	r	$\neg r$	$q \land \neg r$	$p \leftrightarrow (q \land \neg r)$
T	T	T	F	F	F
T	T	F	T	T	T
T	F	T	F	F	F
T	F	F	T	F	F
F	T	T	F	F	T
F	T	F	T	T	F
F	F	T	F	F	T
F	F	F	T	F	T

2. We use the highlighted row in the truth table above to get a truth value of **false**.

3. $p \leftrightarrow (q \land \neg r) \equiv T \leftrightarrow (q \land \neg T) \equiv T \leftrightarrow (q \land F) \equiv T \leftrightarrow F \equiv F$. So, the truth value **can be determined** and it is **false**.

Exercise 1.23: We will show that the truth tables for $p \to q$ and $\neg p \lor q$ are the same.

p	q	$p \to q$	$\neg p$	$\neg p \lor q$
T	T	T	F	T
T	F	F	F	F
F	T	T	T	T
F	F	T	T	T

Exercise 1.25: We will show that the truth tables for $p \to q$ and $\neg q \to \neg p$ are the same.

p	q	$p \to q$	$\neg p$	$\neg q$	$\neg q \to \neg p$
T	T	T	F	F	T
T	F	F	F	T	F
F	T	T	T	F	T
F	F	T	T	T	T

Exercise 1.27: We will show that the truth tables for $\neg(p \lor q)$ and $\neg p \land \neg q$ are the same.

p	q	$\neg p$	$\neg q$	$p \lor q$	$\neg(p \lor q)$	$\neg p \land \neg q$
T	T	F	F	T	F	F
T	F	F	T	T	F	F
F	T	T	F	T	F	F
F	F	T	T	F	T	T

Exercise 1.30:

$$[(\neg p \lor q) \land p] \lor q \equiv [p \land (\neg p \lor q)] \lor q \equiv [(p \land \neg p) \lor (p \land q)] \lor q \equiv [F \lor (p \land q)] \lor q$$
$$\equiv [(p \land q) \lor F] \lor q \equiv (p \land q) \lor q \equiv (q \land p) \lor q \equiv q$$

So, we see that $[(\neg p \lor q) \land p] \lor q$ is logically equivalent to the atomic statement q.

Notes: (1) For the first equivalence, we used the first commutative law.

(2) For the second equivalence, we used the first distributive law.

(3) For the third equivalence, we used the first negation law.

(4) For the fourth equivalence, we used the second commutative law.

(5) For the fifth equivalence, we used the fourth identity law.

(6) For the sixth equivalence, we used the first commutative law.

(7) For the last equivalence, we used the second absorption law.

Exercise 1.32: We will show that the final column of the truth table for $(p \rightarrow q) \leftrightarrow (\neg q \rightarrow \neg p)$ consists of only the truth value T.

p	q	$\neg p$	$\neg q$	$p \rightarrow q$	$\neg q \rightarrow \neg p$	$(p \rightarrow q) \leftrightarrow (\neg p \rightarrow \neg q)$
T	T	F	F	T	T	T
T	F	F	T	F	F	T
F	T	T	F	T	T	T
F	F	T	T	T	T	T

Exercise 1.34: If $p \equiv T$, then $p \wedge \neg p \equiv T \wedge F \equiv F$. If $p \equiv F$, then $p \wedge \neg p \equiv F \wedge T \equiv F$. Since both possible truth assignments of the propositional variable p lead to the statement $p \wedge \neg p$ having truth value F, it follows that $p \wedge \neg p$ is a contradiction.

Lesson 2

Exercise 2.2:

1. 5 elements; p, q, r, s and t.

2. 3 elements; centipede, pencil, and artichoke

3. 4 elements; $2.5, 7.01, 11.3$, and 19.65

4. 6 elements; blue, red, yellow, green, purple, and orange

5. 4 elements; Zeus, Hera, Andromeda, and Hermes

Exercise 2.4:

1. $\{2, 4, 6\}$, $\{4, 2, 6\}$, $\{1, 1, 4, 6\}$, $\{2, 2, 4, 6, 6\}$

2. $\{a, x, t, v, d, b\}$, $\{a, a, x, x, x, t, t, t, t, v, v, v, d, d\}$, $\{t, a, x, d, v, d, x, a, t\}$, $\{t, a, x, x, v, d\}$

3. $\{cat, dog\}$, $\{c, a, t, d, o, g\}$, $\{dog, cat, dog\}$, $\{cat, cat, dog, cat\}$

Exercise 2.6:

1. This is **true** because 5 is an element of Y.

2. This is **false** because z is an element of Y.

3. This is **false** because hawk is **not** an element of Y.

4. This is **false** because 6 is **not** an element (or member) of Y.

5. This is **true** because y is **not** an element of Y.

6. This is **true** because 12 and t are both elements of Y.

7. This is **true** because $0, 5, 12$, and eagle are all elements of Y.

8. This is **false** because x is **not** an element of Y.

Exercise 2.8: $\{1, 3, 5, 7, 9, 11, 13, 15, 17, 19, 21, 23, 25\}$

Exercise 2.10:

1. $2\mathbb{Z} = \{..., -6, -4, -2, 0, 2, 4, 6, ...\}$

2. $2\mathbb{Z} + 1 = \{..., -5, -3, -1, 1, 3, 5, 7, ...\}$

3. $\mathbb{Z}^- = \{..., -6, -5, -4, -3, -2, -1\}$

Exercise 2.12:

1. **No** because the word "hello" contains just two vowels: e and o.

2. **Yes** because the word "cuttlefish" contains the three distinct vowels u, e, and i.

3. **No** because the phrase "gone away" consists of two words and not a single word.

4. **Yes** because the word "existential" contains the three distinct vowels e, i, and a.

5. **No** because the word "sediment" contains just two distinct vowels: e and i. Note that in this example, the word contains three vowels (because the e appears twice), but only two **distinct** vowels.

Exercise 2.14: Note that there are many different correct descriptions for each of the sets given. We provide just one of each here.

1. $\{n \mid n \text{ is an even natural number and } 0 \leq n \leq 12\}$

2. $\{n \mid n \text{ is an odd integer between } -3 \text{ and } 99, \text{ inclusive}\}$

3. $\{n \mid n \text{ is a natural number}\}$

4. $\{n \mid n \text{ is an integer}\}$

5. $\{n \mid n \text{ is an even integer}\}$

Exercise 2.16: $\dfrac{3}{5}, \dfrac{6}{10}$ $\dfrac{-13}{7}, \dfrac{13}{-7}$ $\dfrac{0}{6}, 0, \dfrac{0}{-1}$ $\dfrac{4}{4}, \dfrac{-17}{-17}, 1, \dfrac{1}{1}$ $\dfrac{-5}{-3}, \dfrac{20}{12}$ $\dfrac{10}{-6}, \dfrac{-15}{9}$

Exercise 2.17:

1. **rational**

2. **rational**

3. **irrational**

4. **rational**

5. **irrational**

Exercise 2.18: See the image to the right.

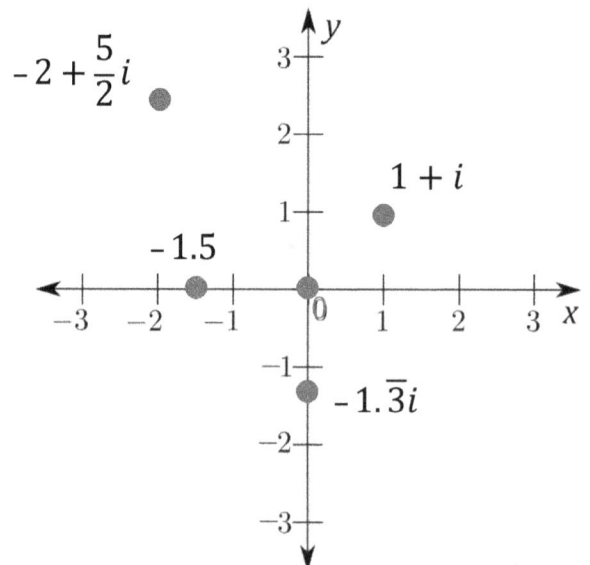

162

Exercise 2.20:

1. $|\{1, 2, 3, \ldots, 50\}| = \mathbf{50}$

2. $|\{c, d, e, f, e, d, c\}| = \mathbf{4}$ (note that $\{c, d, e, f, e, d, c\} = \{c, d, e, f\}$)

3. $|\{\emptyset, \{\emptyset\}\}| = \mathbf{2}$ (the two elements of the set are $\emptyset$ and $\{\emptyset\}$)

4. $|\{n \in \mathbb{N} \mid 126 \leq n \leq 2007\}| = 2007 - 126 + 1 = \mathbf{1882}$ (use the fence-post formula)

5. $\{x, x\} = \{x\}$, $\{x, x, x\} = \{x\}$, and $\{x, \{x, x\}\} = \{x, \{x\}\}$.

   Therefore, $\{x, \{x, x\}, \{x, x, x\}, \{x, \{x, x\}\}\} = \{x, \{x\}, \{x, \{x\}\}\}$.

   So, $\left|\{x, \{x, x\}, \{x, x, x\}, \{x, \{x, x\}\}\}\right| = \left|\{x, \{x\}, \{x, \{x\}\}\}\right| = \mathbf{3}$. (the three elements of the set are $x, \{x\}$, and $\{x, \{x\}\}$)

Exercise 2.22:

1. $\mathbf{\mathit{B} \subseteq \mathit{A}}$. Also, $A \nsubseteq B$ because $y \in A$, whereas $y \notin B$.

2. **Neither**. $A \nsubseteq B$ because $-2 \in A$, whereas $-2 \notin B$. $B \nsubseteq A$ because $1 \in B$, whereas $1 \notin A$.

3. **Both**. $A = \{x, \{x, x\}\} = \{x, \{x\}\}$ and $B = \{\{x\}, x, x\} = \{\{x\}, x\} = \{x, \{x\}\}$. So, $A = B$.

4. $\mathbf{\mathit{B} \subseteq \mathit{A}}$. Also, $A \nsubseteq B$ because $-3 \in A$, whereas $-3 \notin B$.

Exercise 2.24:

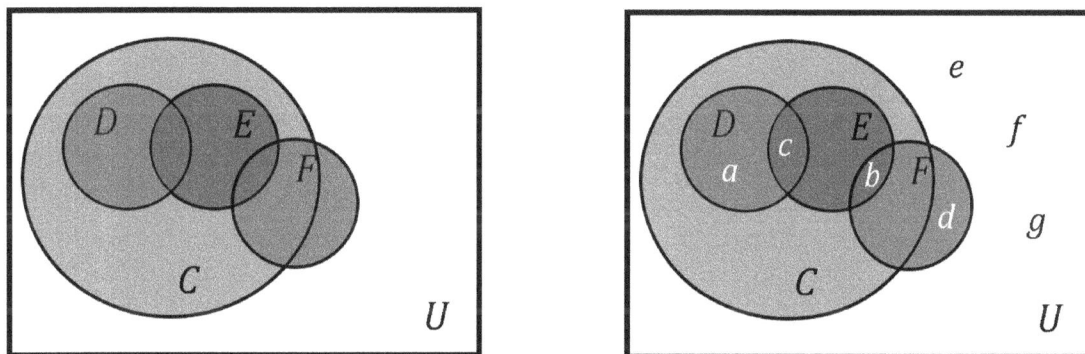

Note: Above are possible Venn diagrams for this problem. The diagram on the left shows the relationship between the sets C, D, E, and F. Notice how D and E are both subsets of C, whereas F is not a subset of C. Also, notice how D and E overlap, E and F overlap, but there is no overlap between D and F (they have no elements in common). The diagram on the right shows the proper placement of the elements. Here, I chose the universal set to be $U = \{a, b, c, d, e, f, g\}$. This choice for the universal set is somewhat arbitrary. Any set containing $\{a, b, c, d\}$ would do.

Exercise 2.26: $\{a, b, c, d\}$ has **16** subsets. We can also say that the cardinality of the power set of $\{a, b, c, d\}$ is 16. That is, $|\mathcal{P}(\{\mathbf{a, b, c, d}\})| = \mathbf{16}$. Below is a tree diagram.

$$\{a,b,c,d\}$$

$$\{a,b,c\} \quad \{a,b,d\} \quad \{a,c,d\} \quad \{b,c,d\}$$

$$\{a,b\} \quad \{a,c\} \quad \{a,d\} \quad \{b,c\} \quad \{b,d\} \quad \{c,d\}$$

$$\{a\} \quad \{b\} \quad \{c\} \quad \{d\}$$

$$\emptyset$$

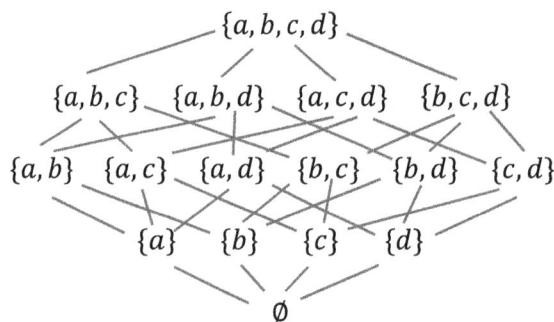

Exercise 2.28: $|\mathcal{P}(X)| = 128 = 2^7$. Therefore, $|X| = \mathbf{7}$.

Exercise 2.30:

1. $A \cup B = \{a, b, c, \Delta, \delta, \gamma\}$.

2. $A \cap B = \{b, \delta\}$.

3. $A \,\Delta\, B = (A \setminus B) \cup (B \setminus A) = \{a, \Delta\} \cup \{c, \gamma\} = \{a, c, \Delta, \gamma\}$

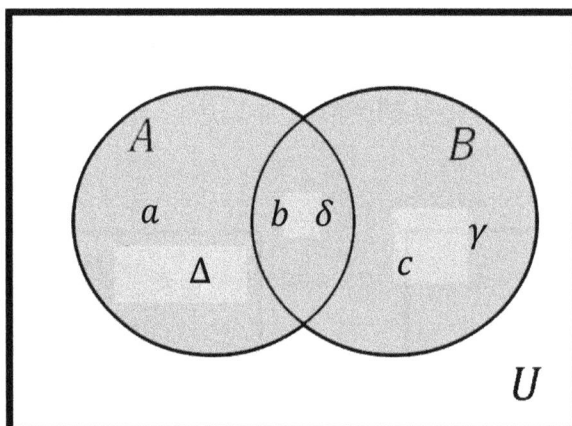

Exercise 2.33:

1. **No.** For example, $1 + 2i \in A$ because $1, 2 \in \mathbb{Z}$, but $1 + 2i \notin B$ because $1 \in \mathbb{Q}$.

2. **No.** For example, $0.1010010001\ldots + 3i \in B$ because $0.1010010001\ldots \notin \mathbb{Q}$, but $0.1010010001\ldots + 3i \notin A$ because $0.1010010001\ldots \notin \mathbb{Z}$.

3. **Yes.** To see this, suppose that $a + bi \in A \cap B$. Then $a + bi \in A$, and so, $a \in \mathbb{Z}$. Since $\mathbb{Z} \subseteq \mathbb{Q}$, $a \in \mathbb{Q}$. Also, $a + bi \in B$. So, $a \notin \mathbb{Q}$. Since we cannot have both $a \in \mathbb{Q}$ and $a \notin \mathbb{Q}$, we must have $A \cap B = \emptyset$.

Exercise 2.34:

1. $A = \{0\}, B = \{1\}$. Then $A \cup B = \{0, 1\}$. Since $1 \in A \cup B$ and $1 \notin A$, $A \cup B \nsubseteq A$.

2. $A = \{0, 1\}, B = \{1, 2\}$. Then $A \cap B = \{1\}$. Since $0 \in A$ and $0 \notin A \cap B$, $A \nsubseteq A \cap B$.

3. $A = \{0, 1\}, B = \{0\}$. Then $B \subseteq A$, but $A \cup B = \{0, 1\} \neq \{0\} = B$.

4. $A = \{0, 1\}, B = \{0\}$. Then $B \subseteq A$, but $A \setminus B = \{1\} \neq \emptyset$.

Lesson 3

Exercise 3.2:

1. **Yes**

2. **No**

3. **No**

4. **Yes**

·	−1	0	1
−1	1	0	−1
0	0	0	0
1	−1	0	1

−	−1	0	1
−1	0	−1	**−2**
0	1	0	−1
1	**2**	1	0

+	−1	0	1
−1	**−2**	−1	0
0	−1	0	1
1	0	1	**2**

⋆	−1	0	1
−1	−1	−1	−1
0	0	0	0
1	1	1	1

Exercise 3.4:

1. **Closed**

2. **Not closed**

3. **Not closed**

4. **Closed**

Exercise 3.6:

1. **Yes**

2. **Yes**

3. $2 \circ 3 = \mathbf{3}$

4. $3 \circ 2 = \mathbf{1}$

∘	0	1	2	3
0	0	1	2	3
1	1	0	1	3
2	2	2	0	3
3	3	2	1	0

∘	0	1	2	3
0	0	1	2	3
1	1	0	1	3
2	2	2	0	3
3	3	2	1	0

Exercise 3.8:

1. **Yes**

2. **No:** For example, let $a = b = 1$. Then $a, b \in 2\mathbb{N} + 1$ (because a and b are odd), but $a + b = 2$ and $2 \notin 2\mathbb{N} + 1$ (because 2 is even).

3. **Yes**

4. **Yes**

Exercise 3.10:

1. **Yes.** Indeed, $(x \star y) \star z = a$ and $x \star (y \star z) = a$ for all $x, y, z \in S$.

2. **No.** For example, $(b \star a) \star b = a \star b = b$, whereas $b \star (a \star b) = b \star b = a$.

3. **No.** For example, $(a \star a) \star b = b \star b = a$, whereas $a \star (a \star b) = a \star b = b$.

4. **Yes.** The following computations verify this:

$$
\begin{aligned}
(a \star a) \star a &= a \star a = a & \qquad a \star (a \star a) &= a \star a = a \\
(a \star a) \star b &= a \star b = a & a \star (a \star b) &= a \star a = a \\
(a \star b) \star a &= a \star a = a & a \star (b \star a) &= a \star b = a \\
(a \star b) \star b &= a \star b = a & a \star (b \star b) &= a \star b = a \\
(b \star a) \star a &= b \star a = b & b \star (a \star a) &= b \star a = b \\
(b \star a) \star b &= b \star b = b & b \star (a \star b) &= b \star a = b \\
(b \star b) \star a &= b \star a = b & b \star (b \star a) &= b \star b = b \\
(b \star b) \star b &= b \star b = b & b \star (b \star b) &= b \star b = b
\end{aligned}
$$

Exercise 3.12:

1. **Yes.** Indeed, $(a \star b) \star c = a \star c = a = a \star (b \star c)$ for all $a, b, c \in \mathbb{N}$.

2. **No.** For example, $(2 \star 1) \star 3 = (2^1)^3 = 2^3 = 8$, whereas $2 \star (1 \star 3) = 2^{(1^3)} = 2^1 = 2$.

3. **Yes.** There are 6 cases to consider:

Case 1 ($a \leq b \leq c$):
$$
\begin{aligned}
(a \star b) \star c &= \min\{a, b\} \star c = a \star c = \min\{a, c\} = a \\
a \star (b \star c) &= a \star \min\{b, c\} = a \star b = \min\{a, b\} = a
\end{aligned}
$$

Case 2 ($a \leq c \leq b$):
$$
\begin{aligned}
(a \star b) \star c &= \min\{a, b\} \star c = a \star c = \min\{a, c\} = a \\
a \star (b \star c) &= a \star \min\{b, c\} = a \star c = \min\{a, c\} = a
\end{aligned}
$$

Case 3 ($b \leq a \leq c$):
$$
\begin{aligned}
(a \star b) \star c &= \min\{a, b\} \star c = b \star c = \min\{b, c\} = b \\
a \star (b \star c) &= a \star \min\{b, c\} = a \star b = \min\{a, b\} = b
\end{aligned}
$$

Case 4 ($b \leq c \leq a$):
$$
\begin{aligned}
(a \star b) \star c &= \min\{a, b\} \star c = b \star c = \min\{b, c\} = b \\
a \star (b \star c) &= a \star \min\{b, c\} = a \star b = \min\{a, b\} = b
\end{aligned}
$$

Case 5 ($c \leq a \leq b$):
$$
\begin{aligned}
(a \star b) \star c &= \min\{a, b\} \star c = a \star c = \min\{a, c\} = c \\
a \star (b \star c) &= a \star \min\{b, c\} = a \star c = \min\{a, c\} = c
\end{aligned}
$$

Case 6 ($c \leq b \leq a$):
$$
\begin{aligned}
(a \star b) \star c &= \min\{a, b\} \star c = b \star c = \min\{b, c\} = c \\
a \star (b \star c) &= a \star \min\{b, c\} = a \star c = \min\{a, c\} = c
\end{aligned}
$$

Exercise 3.14:

1. **No.** For example, $2 \star 5 = 2$, whereas $5 \star 2 = 5$.

2. **Yes.** There are 2 cases to consider:

   Case 1 ($a \leq b$): $a \star b = \min\{a, b\} = a = \min\{b, a\} = b \star a$

   Case 2 ($b \leq a$): $a \star b = \min\{a, b\} = b = \min\{b, a\} = b \star a$

Exercise 3.16:

1. **No.** a does not even appear as an output in the multiplication table.

2. **Yes.** b is an identity. We can see this visually as follows:

$\star$	a	b	c
a	a	a	a
b	a	b	c
c	a	c	b

$\star$	a	b	c
a	a	a	a
b	a	b	c
c	a	c	b

Exercise 3.18:

1. $(2\mathbb{N}, +)$ **is a commutative monoid** with identity 0.

2. $(\mathbb{N}, \star)$ **is a noncommutative semigroup** by part 1 of Exercise 3.12 and part 1 of exercise 3.14. $(\mathbb{N}, \star)$ **is not a monoid**. To see this, let $a \in \mathbb{N}$. Then $a + 1 \in \mathbb{N}$ and $a \star (a + 1) = a \neq a + 1$. This shows that a is not an identity. Since a was an arbitrary element of $\mathbb{N}$, we showed that there is no identity.

3. $(\mathbb{Z}, \star)$ **is a commutative semigroup** by part 3 of Exercise 3.12 and part 2 of exercise 3.14. $(\mathbb{Z}, \star)$ **is not a monoid** by the same reasoning as in 2 above.

4. $(2\mathbb{Z}, \cdot)$ **is a commutative semigroup** by part 3 of Example 3.13. $(2\mathbb{Z}, \cdot)$ **is not a monoid**. The identity element of $(\mathbb{Z}, \cdot)$ is 1, and this element is missing from $(2\mathbb{Z}, \cdot)$.

5. $(\mathcal{P}(A), \cap)$ **is a commutative monoid**. If $X, Y \in \mathcal{P}(A)$, then every element of X is in A. Since every element of $X \cap Y$ is an element of X, it follows that every element of $X \cap Y$ is in A. So, $X \cap Y \in \mathcal{P}(A)$. This shows that $\cap$ is a binary operation on $\mathcal{P}(A)$. $\cap$ is associative and commutative in $\mathcal{P}(A)$. Finally, A is an identity for $(\mathcal{P}(A), \cap)$ because if $X \in \mathcal{P}(A)$, then $X \cap A = X$ and $A \cap X = X$.

Exercise 3.19:

1. $e \star f = f$ because e **is an identity**.

2. $e \star f = e$ because f **is an identity**.

Exercise 3.21:

1. **Yes.** $(2\mathbb{Z}, +)$ is a commutative group with identity 0. The inverse of $2k$ is $-2k$.

2. **No.** The natural number 2 has no additive inverse in $2\mathbb{N}$.

3. **No.** The integer 2 has no multiplicative inverse in $\mathbb{Z}$.

4. **No.** A has no inverse in $\mathcal{P}(A)$. Indeed, if $B \in \mathcal{P}(A)$, then since $B \subseteq A$, $A \cup B = A \neq \emptyset$.

Exercise 3.23:

1. Let $\frac{a}{b}, \frac{c}{d} \in \mathbb{Q}$. Then $a, b, c, d \in \mathbb{Z}$ and $b, d \neq 0$. Since $\mathbb{Z}$ is closed under multiplication, $ad, bc, bd \in \mathbb{Z}$. Since $\mathbb{Z}$ is closed under addition, $ad + bc \in \mathbb{Z}$. Since $b, d \neq 0$, $bd \neq 0$. Therefore, $\frac{a}{b} + \frac{c}{d} = \frac{ad+bc}{bd} \in \mathbb{Q}$.

2. Let $\frac{a}{b}, \frac{c}{d} \in \mathbb{Q}^*$. Then $a, b, c, d \in \mathbb{Z}^*$. Since $\mathbb{Z}^*$ is closed under multiplication, $ac, bd \in \mathbb{Z}^*$. Therefore, $\frac{a}{b} \cdot \frac{c}{d} = \frac{ac}{bd} \in \mathbb{Q}^*$.

3. $\frac{-a}{b}$

4. $\frac{b}{a}$

Exercise 3.25:

1. **Yes**. Inverse pairs: $\{0, 0\}, \{1, 5\}, \{2, 4\}, \{3, 3\}$

2. **Yes**. Inverse pairs: $\{1, 1\}, \{2, 4\}, \{3, 5\}, \{6, 6\}$

3. **No** because $2 \cdot 5 = 0$.

Exercise 3.27:

1. $a \star b = e$ because b **is an inverse of** a.

2. $c \star a = e$ because c **is an inverse of** a.

3. $b = e \star b = (c \star a) \star b = c \star (a \star b) = c \star e = c$.

Note: In part 3 above, we used the following properties: For the first and fifth equalities, we used the fact that e is an identity in G. For the second equality, we used the fact that $c \star a = e$ and for the fourth equality, we used the fact that $a \star b = e$. For the third equality, we used the associativity of $\star$ in G.

Exercise 3.30:

1. **Yes**

2. **Yes**

3. **No** because $0 \notin \mathbb{Z}^+$.

4. **Yes**

Exercise 3.31:

$$0 \cdot a = 0 \cdot a + 0 = 0 \cdot a + (0 \cdot a - 0 \cdot a) = (0 \cdot a + 0 \cdot a) - 0 \cdot a$$
$$= (0 + 0)a - 0 \cdot a = 0 \cdot a - 0 \cdot a = 0.$$

Exercise 3.34:

1. **Yes** because 11 is prime.

2. **No** because 15 is not prime ($15 = 3 \cdot 5$).

Lesson 4

Exercise 4.2:

1. $b = \mathbf{3}$. So, $6 = 2 \cdot 3$. This equation shows that 6 is even because 3 is an integer.

2. $b = -\mathbf{1}$. So, $-2 = 2(-1)$. This equation shows that -2 is even because -1 is an integer.

3. $b = \frac{\mathbf{5}}{\mathbf{2}}$. So, $5 = 2 \cdot \frac{5}{2}$. This equation **does not** show that 5 is even because $\frac{5}{2}$ is **not** an integer.

4. $b = \mathbf{0}$. So, $0 = 2 \cdot 0$. This equation shows that 0 is even because 0 is an integer.

5. $b = -\frac{\mathbf{1}}{\mathbf{2}}$. So, $-1 = 2\left(-\frac{1}{2}\right)$. This equation **does not** show that -1 is even because $-\frac{1}{2}$ is **not** an integer.

Exercise 4.4: Let m and n be even integers. Since m is even, there is an integer j such that $m = 2j$. Since n is even, there is an integer k such that $n = 2k$. So, we have

$$mn = (2j)(2k) = 2\big(j(2k)\big).$$

For the first equality, we made two substitutions.

For the second equality, we used the associativity of multiplication in $\mathbb{Z}$.

Note that $j(2k)$ is an integer because the set of integers is closed under multiplication.

It follows that mn is even.

Since m and n were arbitrary even integers, we have verified Even Integer Fact 2.

Exercise 4.6:

1. $b = \mathbf{4}$. So, $12 = 3 \cdot 4$.

2. $b = \mathbf{3}$. So, $12 = 4 \cdot 3$.

3. $b = -\mathbf{7}$. So, $-77 = 11(-7)$.

4. $b = \mathbf{1}$. So, $13 = 13 \cdot 1$.

5. $b = -\mathbf{4}$. So, $24 = (-6)(-4)$.

Exercise 4.8:

1. $2, 3, 5, 7, 11, 13, 17, 19, 23, 29, 31, 37, 41, 43, 47, 53, 59, 61, 67, 71$

2. $4, 6, 8, 9, 10, 12, 14, 15, 16, 18, 20, 21, 22, 24, 25, 26, 27, 28, 30, 32$

3. **Composite** because $91 = 7 \cdot 13$.

Exercise 4.9: Let n be an integer greater than 1. By Prime Number Fact 2, n can be written as a product of prime numbers. Let p be any of the prime numbers in that product. Then p is a prime factor of n.

Exercise 4.11:

1. **Factorizations that are not prime:** $1 \cdot 12, 2 \cdot 6, 3 \cdot 4$ **Prime factorization:** $2 \cdot 2 \cdot 3$

2. **Factorizations that are not prime:** $1 \cdot 18, 2 \cdot 9, 3 \cdot 6$ **Prime factorization:** $2 \cdot 3 \cdot 3$

3. **Factorizations that are not prime:** $1 \cdot 100, 2 \cdot 50, 4 \cdot 25$ **Prime factorization:** $2 \cdot 2 \cdot 5 \cdot 5$

Notes: (1) We have listed all factorizations of 12 and 18, up to the order in which the factors are written. In other words, all factorizations that we didn't list contain the same factors as one of the factorizations listed, but written in a different order.

For example, we are considering $12 \cdot 1$ to be the same factorization as $1 \cdot 12$ and we are considering $2 \cdot 3 \cdot 2$ to be the same factorization as $2 \cdot 2 \cdot 3$.

(2) There are additional factorizations of 100 that we have not listed. For example, $5 \cdot 20, 10 \cdot 10$, and $4 \cdot 5 \cdot 5$. Can you find all the others?

Exercise 4.13:

1. **composite** $119 = 7 \cdot 17$

2. **composite** $437 = 19 \cdot 23$

3. **prime**

4. **prime**

5. **composite** $1457 = 31 \cdot 47$

Exercise 4.15:

6. $1 \cdot 2 \cdot 3 \cdot 4 \cdot 5 = \mathbf{120}$.

7. $1 \cdot 2 \cdot 3 \cdot 4 \cdot 5 \cdot 6 = \mathbf{720}$.

8. $1 \cdot 2 \cdot 3 \cdot 4 \cdot 5 \cdot 6 \cdot 7 = \mathbf{5040}$.

Exercise 4.16:

1. The factors of $5! = 120$ are $1, 2, 3, 4, 5, 6, 8, 10, 12, 15, 20, 24, 30, 40, 60$, and 120

2. $6! = 1 \cdot 2 \cdot 3 \cdot 4 \cdot 5 \cdot 6 = 2^4 \cdot 3^2 \cdot 5$. So, the number of factors is $5 \cdot 3 \cdot 2 = \mathbf{30.}$

Exercise 4.17: Let $M = n! + 1$. By the previous paragraph, M is **not** divisible by $1, 2, 3, ..., n$. By Exercise 4.9, M has a prime factor q. Since q cannot be $1, 2, 3, ...,$ or n, we must have $q > n$.

Exercise 4.20:

1. **Even.** $46 = 2 \cdot 23$. The quotient is **23** and the remainder is **0**.

2. **Odd.** $97 = 2 \cdot 48 + 1$. The quotient is **48** and the remainder is **1**.

3. **Even.** $-38 = 2(-19)$. The quotient is **-19** and the remainder is **0**.

4. **Odd.** $-51 = 2(-26) + 1$. The quotient is **-26** and the remainder is **1**.

Exercise 4.22:

1. $23 = 7 \cdot 3 + 2$. So, the quotient is **3** and the remainder is **2**.

2. $66 = 11 \cdot 6 + 0 = 11 \cdot 6$. So, the quotient is **6** and the remainder is **0**.

3. $-49 = 8(-7) + 7$. So, the quotient is **-7** and the remainder is **7**.

Exercise 4.24:

1. $\gcd(17, 51) = \mathbf{17}$ because 17 is a divisor of 51. So, 17 and 51 are **not relatively prime**.

2. The divisors of 75 are $\pm 1, \pm 3, \pm 5, \pm 15, \pm 25, \pm 75$. The divisors of 90 are $\pm 1, \pm 2, \pm 3, \pm 5, \pm 6, \pm 9, \pm 10, \pm 15 \pm 18 \pm 30 \pm 45, \pm 90$. The common divisors of 75 and 90 are $\pm 1, \pm 3, \pm 5, \pm 15$. So, $\gcd(75, 90) = \mathbf{15}$. Therefore, 75 and 90 are **not relatively prime**.

3. The divisors of 19 are $\pm 1, \pm 19$. The divisors of 31 are $\pm 1, \pm 31$. The common divisors of 19 and 31 are ± 1. So, $\gcd(19, 31) = \mathbf{1}$. 19 and 31 are **relatively prime**.

4. The divisors of 170 are $\pm 1, \pm 2, \pm 5, \pm 10, \pm 17, \pm 34 \pm 85, \pm 170$. The divisors of 483 are $\pm 1, \pm 3, \pm 7, \pm 21, \pm 23, \pm 69 \pm 161, \pm 483$. The common divisors of 170 and 483 are ± 1. So, $\gcd(170, 483) = \mathbf{1}$. 170 and 483 are **relatively prime**.

Exercise 4.26:

1. $\mathrm{lcm}(17, 51) = \mathbf{51}$ because 51 is a multiple of 17.

2. The positive multiples of 90 are $90, 180, 270, 360, \mathbf{450}, ...$ Since 450 is the first number in this list that is also a multiple of 75, $\mathrm{lcm}(75, 90) = \mathbf{450}$.

3. $\mathrm{lcm}(19, 31) = 19 \cdot 31 = \mathbf{589}$ (because 19 and 31 are relatively prime).

4. $\mathrm{lcm}(170, 483) = 170 \cdot 483 = \mathbf{82,110}$. (because 170 and 483 are relatively prime).

Exercise 4.28:

1. $\gcd(14, 21, 77) = \mathbf{7}$ and $\mathrm{lcm}(14, 21, 77) = \mathbf{462}$. The integers $14, 21$, and 77 are **neither** mutually nor pairwise relatively prime.

2. $\gcd(2, 3, 5, 7) = \mathbf{1}$ and $\mathrm{lcm}(2, 3, 5, 7) = 2 \cdot 3 \cdot 5 \cdot 7 = \mathbf{210}$. The integers $2, 3, 5$, and 7 are **both** mutually and pairwise relatively prime.

3. $\gcd(55, 85, 187) = \mathbf{1}$ and $\mathrm{lcm}(55, 85, 187) = \mathbf{935}$. The integers $55, 85$, and 187 are **mutually relatively prime**, but **not** pairwise relatively prime.

4. $\gcd(300, 450, 1470) = \mathbf{30}$ and $\mathrm{lcm}(300, 450, 1470) = \mathbf{44,100}$. The integers $300, 85$, and 187 are **neither** mutually nor pairwise relatively prime.

Exercise 4.31:

1. $a = 2^3 \cdot 3^2 \cdot 5^0, b = 2^1 \cdot 3^1 \cdot 5^2$. Therefore,

$$\gcd(a, b) = 2^1 \cdot 3^1 \cdot 5^0 = \mathbf{2 \cdot 3} \qquad \mathrm{lcm}(a, b) = \mathbf{2^3 \cdot 3^2 \cdot 5^2}.$$

2. $a = 2^5 \cdot 3^0 \cdot 5^0 \cdot 7^3, b = 2^0 \cdot 3^2 \cdot 5^3 \cdot 7^0$. Therefore,

$$\gcd(a, b) = 2^0 \cdot 3^0 \cdot 5^0 \cdot 7^0 = \mathbf{1} \qquad \mathrm{lcm}(a, b) = \mathbf{2^5 \cdot 3^2 \cdot 5^3 \cdot 7^3}.$$

3. $a = 2^5 \cdot 3^0 \cdot 5^0 \cdot 7^0 \cdot 11^2 \cdot 13^0, b = 2^0 \cdot 3^2 \cdot 5^0 \cdot 7^0 \cdot 11^1 \cdot 13^2$. Therefore,

$$\gcd(a, b) = 2^0 \cdot 3^0 \cdot 5^0 \cdot 7^0 \cdot 11^1 \cdot 13^0 = \mathbf{11}$$
$$\text{lcm}(a, b) = 2^5 \cdot 3^2 \cdot 5^0 \cdot 7^0 \cdot 11^2 \cdot 13^2 = \mathbf{2^5 \cdot 3^2 \cdot 11^2 \cdot 13^2}.$$

4. $a = 2^1 \cdot 3^0 \cdot 5^0 \cdot 7^0 \cdot 11^0 \cdot 13^0 \cdot 17^1 \cdot 19^2, b = 2^0 \cdot 3^0 \cdot 5^2 \cdot 7^0 \cdot 11^0 \cdot 13^0 \cdot 17^3 \cdot 19^0$. Therefore,

$$\gcd(a, b) = 2^0 \cdot 3^0 \cdot 5^0 \cdot 7^0 \cdot 11^0 \cdot 13^0 \cdot 17^1 \cdot 19^0 = \mathbf{17}$$
$$\text{lcm}(a, b) = 2^1 \cdot 3^0 \cdot 5^2 \cdot 7^0 \cdot 11^0 \cdot 13^0 \cdot 17^3 \cdot 19^2 = \mathbf{2 \cdot 5^2 \cdot 17^3 \cdot 19^2}.$$

Exercise 4.33:

1. $12 = 1 \cdot 12 + 0 \cdot 18.$

2. $18 = 0 \cdot 12 + 1 \cdot 18$

3. $6 = 2 \cdot 12 - 1 \cdot 18$

4. $36 = 12 \cdot 12 - 6 \cdot 18$

5. **Not possible** because $\gcd(12, 18) = 6$ and $3 < 6$.

Exercise 4.35:

$$1040 = 305 \cdot 3 + 125$$
$$305 = 125 \cdot 2 + 55$$
$$125 = 55 \cdot 2 + 15$$
$$55 = 15 \cdot 3 + 10$$
$$15 = 10 \cdot 1 + \mathbf{5}$$
$$10 = 5 \cdot 2 + \mathbf{0}$$

So, $\gcd(305, 1040) = \mathbf{5}$.

Exercise 4.37: We start with the second to last line of the solution to Exercise 4.35 (line 5):

$$15 = 10 \cdot 1 + 5.$$

We solve this equation for 5 to get $5 = 15 - 1 \cdot 10$.

Working backwards, we next look at line 4: $55 = 15 \cdot 3 + 10$. We solve this equation for 10 and then substitute into the previous equation: $10 = 55 - 15 \cdot 3$. After substituting, we get

$$5 = 15 - 1 \cdot 10 = 15 - 1(55 - 15 \cdot 3).$$

We then distribute and group all the 15's together and all the 55's together. So, we have

$$5 = 15 - 1 \cdot 10 = 15 - 1(55 - 15 \cdot 3) = 15 - 1 \cdot 55 + 3 \cdot 15 = 4 \cdot 15 - 1 \cdot 55.$$

Line 3 is next: $125 = 55 \cdot 2 + 15$. We solve this equation for 15 to get $15 = 125 - 2 \cdot 55$. And once again we now substitute into the previous equation to get

$$5 = 4 \cdot 15 - 1 \cdot 55 = 4(125 - 2 \cdot 55) - 1 \cdot 55 = 4 \cdot 125 - 8 \cdot 55 - 1 \cdot 55 = 4 \cdot 125 - 9 \cdot 55.$$

Let's go to line 2: $305 = 125 \cdot 2 + 55$. We solve this equation for 55 to get $55 = 305 - 2 \cdot 125$. Substituting into the previous equation gives us

$$5 = 4 \cdot 125 - 9 \cdot 55 = 4 \cdot 125 - 9(305 - 2 \cdot 125)$$
$$= 4 \cdot 125 - 9 \cdot 305 + 18 \cdot 125 = 22 \cdot 125 - 9 \cdot 305.$$

And finally line 1: $1040 = 305 \cdot 3 + 125$. Solving this equation for 125 gives us $125 = 1040 - 3 \cdot 305$. Substituting into the previous equation gives

$$5 = 22 \cdot 125 - 9 \cdot 305 = 22(1040 - 3 \cdot 305) - 9 \cdot 305$$
$$= 22 \cdot 1040 - 66 \cdot 305 - 9 \cdot 305 = 22 \cdot 1040 - 75 \cdot 305.$$

So, we see that $\gcd(305, 1040) = 5 = 22 \cdot 1040 - 75 \cdot 305 = \mathbf{-75 \cdot 305 + 22 \cdot 1040}$.

Note: With a little practice, the computations above can be done fairly quickly. Here is what a quicker computation might look like:

$$5 = 15 - 1 \cdot 10 = 15 - 1 \cdot (55 - 15 \cdot 3) = 4 \cdot 15 - 1 \cdot 55$$
$$= 4(125 - 55 \cdot 2) - 1 \cdot 55 = 4 \cdot 125 - 9 \cdot 55$$
$$= 4 \cdot 125 - 9(305 - 125 \cdot 2) = 22 \cdot 125 - 9 \cdot 305$$
$$= 22(1040 - 305 \cdot 3) - 9 \cdot 305 = 22 \cdot 1040 - 75 \cdot 305$$

So, $5 = \gcd(305, 1040) = -75 \cdot 305 + 22 \cdot 1040$.

Lesson 5

Exercise 5.2:

1. **False**.

2. **True**.

3. **True**.

4. **False**.

Exercise 5.4:

1. **No**. For example, $1 \not> 1$.

2. **No**. For example, $2 > 1$, but $1 \not> 2$.

3. **Yes**.

4. **Yes**.

5. **Yes**. This is true vacuously because $a > b$ and $b > a$ cannot occur simultaneously.

6. **Yes**.

Exercise 5.6:

1. **No** because $\leq$ is **not** trichotomous by part 3 of Example 5.3.

2. **No** because $\leq$ is **not** trichotomous (for example, $0 \leq 0$ and $0 = 0$ both hold).

3. **Yes**.

4. **Yes**.

5. **No** because $\subset$ is **not** trichotomous: if $a, b \in A$ are distinct, then $\{a\} \not\subset \{b\}$, $\{b\} \not\subset \{a\}$, and $\{a\} \neq \{b\}$.

Exercise 5.7:

1. If $a \in A$, then $a \leq a$ and $a = a$.

2. Let $a \in A$. Since $a = a$, by definition we have $a \leq a$.

3. Let $a, b \in A$ with $a \leq b$ and $b \leq a$. If $a \neq b$, then $a < b$ and $b < a$. But this cannot happen because $<$ is trichotomous on A. So, $a = b$.

4. Let $a, b, c \in A$ with $a \leq b$ and $b \leq c$. If $a = b$, then we have $a \leq c$ by direct substitution. Similarly, if $b = c$, we have $a \leq c$ by direct substitution. If $a < b$ and $b < c$, then $a < c$ because $<$ is transitive. It follows that $a \leq c$.

5. Let $a, b \in A$. If $a < b$ or $a = b$, then $a \leq b$. Otherwise, $b < a$, in which case we have $b \leq a$.

Exercise 5.10:

1. $c^2 = 7^2 + 24^2 = 49 + 576 = 625$. Therefore, $c = \mathbf{25}$.

2. $15^2 = 12^2 + b^2$. So, $225 = 144 + b^2$, and thus, $b^2 = 225 - 144 = 81$. Therefore, $\boldsymbol{b = 9}$.

Exercise 5.12:

1. **Not bounded above**

2. **Bounded above**, Least upper bound $= -\mathbf{1}$

3. **Bounded above**, Least upper bound $= \mathbf{12}$

4. **Bounded above**, Least upper bound $= \mathbf{1}$

5. **Bounded above**, Least upper bound $= \mathbf{135}$

6. **Not bounded above**

7. **Bounded above**, Least upper bound $= -\mathbf{500}$

8. **Bounded above**, Least upper bound **does not exist**

Exercise 5.14:

1. **Not bounded below**

2. **Bounded below**, Greatest lower bound $= \mathbf{0}$

3. **Bounded below**, Greatest lower bound $= \mathbf{0}$

4. **Bounded below**, Greatest lower bound $= 0$

5. **Bounded below**, Greatest lower bound $= -7$

6. **Bounded below**, Greatest lower bound $= -500$

7. **Not bounded below**

8. **Bounded below**, Greatest lower bound **does not exist**

Exercise 5.16:

1. **Unbounded**

2. **Unbounded**

3. **Bounded**

4. **Bounded**

5. **Bounded**

6. **Unbounded**

7. **Unbounded**

8. **Bounded**

Lesson 6

Exercise 6.2:

1. **Interval**.

2. **Not an interval**. $0, 1 \in \mathbb{Z}$, $0 < \frac{1}{2} < 1$, but $\frac{1}{2} \notin \mathbb{Z}$.

3. **Not an interval**. $\frac{1}{2}, 1 \in G$, $\frac{1}{2} < \frac{3}{4} < 1$, but $\frac{3}{4} \notin G$.

4. **Interval**.

5. **Interval**.

6. **Not an interval**. $0, 1 \in \mathbb{Q}$, $0 < 0.01001000100001 \dots < 1$, but $0.01001000100001 \dots \notin \mathbb{Q}$.

Exercise 6.4:

Exercise 6.6:

1. $C \cup D = (-\infty, 3]$

2. $C \cap D = (-1, 2]$

3. $C \setminus D = (-\infty, -1]$

4. $D \setminus C = (2, 3]$

5. $C \, \Delta \, D = (-\infty, -1] \cup (2, 3]$

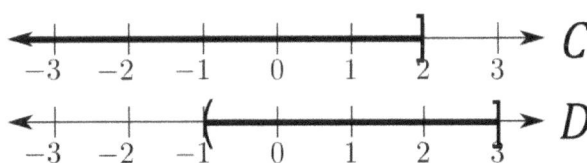

175

Exercise 6.8:

1. $\cup X = \{a, b, c, d, e, x, y, z\}$ $\cap X = \{b\}$

2. $\cup X = [-17, \infty)$ $\cap X = (2, 3)$

3. $\cup X = \mathbb{Z}$ $\cap X = \{0\}$

Exercise 6.10:

1. **Open**

2. **Not open**

3. **Open**

4. **Not open**

5. **Open**

6. **Open** (the definition of open is vacuously satisfied)

Exercise 6.12:

1. **Open** by Open Set Fact 1

2. **Not open** by Problem 42 in Problem Set 5

3. **Not open** by the Density Property of $\mathbb{R}$ (Real Number Fact 2)

4. **Open** by Open Set Fact 1

5. **Open** by Open Set Fact 3

6. **Not open** (This set is equal to $(0, 1]$.)

Exercise 6.14:

1. **Not closed**

2. **Closed**

3. **Not closed**

4. **Closed**

5. **Closed** (because $\mathbb{R} \setminus \mathbb{R} = \emptyset$ is open)

6. **Closed** (because $\mathbb{R} \setminus \emptyset = \mathbb{R}$ is open)

Lesson 7

Exercise 7.2:

1. **2**

2. $-8 - i$

3. $9 + 2i$

Exercise 7.4:

1. $(1+1) + (-1+1)i = \mathbf{2}$

2. $(15+2) + (-3+10)i = \mathbf{17+7i}$

3. $(0+5i)(9-3i) = (0+15) + (0+45)i = \mathbf{15+45i}$

Exercise 7.5:

1. Additive inverse: $-\mathbf{1-i}$ Multiplicative inverse: $\frac{1}{2} - \frac{1}{2}i$

2. Additive inverse: $-\mathbf{2+3i}$ Multiplicative inverse: $\frac{2}{\sqrt{13}} + \frac{3}{\sqrt{13}}i$

3. Additive inverse: $-\mathbf{5i}$ Multiplicative inverse: $-\frac{1}{5}i$

4. Additive inverse: $-\mathbf{7}$ Multiplicative inverse: $\frac{1}{7}$

5. Additive inverse: $-\sqrt{2} + \sqrt{5}i$ Multiplicative inverse: $\frac{\sqrt{2}}{7} + \frac{\sqrt{5}}{7}i$

Exercise 7.7:

1. $\mathbf{2i}$

2. $\mathbf{2-3i}$

3. $\mathbf{-9+8i}$

Exercise 7.8:

1. $\mathbf{1-i}$

2. $\mathbf{2+3i}$

3. $-\mathbf{5i}$

4. $\mathbf{7}$

5. $\sqrt{2} + \sqrt{5}i$

Exercise 7.10:

1. $\frac{1+i}{1-i} = \frac{(1+i)(1+i)}{(1-i)(1+i)} = \frac{2i}{2} = \mathbf{i}$

2. $\frac{-3-2i}{-5+i} = \frac{(-3-2i)(-5-i)}{(-5+i)(-5-i)} = \frac{13+13i}{26} = \frac{13}{26} + \frac{13}{26}i = \frac{1}{2} + \frac{1}{2}i$

3. $\frac{5i}{9-3i} = \frac{5i(9+3i)}{(9-3i)(9+3i)} = \frac{-15+45i}{90} = -\frac{15}{90} + \frac{45}{90}i = -\frac{1}{6} + \frac{1}{2}i$

Exercise 7.12: We are looking for a complex number $a+bi$ such that $(a+bi)^2 = i$. In other words, we need $(a+bi)(a+bi) = (a^2 - b^2) + 2abi$ to equal $i = 0 + 1i$. So, we have $a^2 - b^2 = 0$ and $2ab = 1$. The first equation is equivalent to $(a+b)(a-b) = 0$, so that $a+b = 0$ or $a-b = 0$. It follows that $a = -b$ or $a = b$.

If $a = -b$, then $2ab = 2(-b)b = -2b^2$. So, $-2b^2 = 1$, or equivalently, $b^2 = -1$. Since the square of a real number must be positive, this is impossible.

If $a = b$, then $2ab = 2b^2$. So, $2b^2 = 1$, or equivalently, $b^2 = \frac{1}{2}$. Therefore, $b = \pm\frac{1}{\sqrt{2}}$. So, $a = \pm\frac{1}{\sqrt{2}}$.

It follows that the two square roots of i are $\frac{1}{\sqrt{2}} + \frac{1}{\sqrt{2}}i$ and $-\frac{1}{\sqrt{2}} - \frac{1}{\sqrt{2}}i$.

Exercise 7.14:

1. $|12 - 5i| = \sqrt{12^2 + (-5)^2} = \sqrt{144 + 25} = \sqrt{169} = \mathbf{13}$.

2. $\left|-\sqrt{7}i\right| = \left|0 - \sqrt{7}i\right| = \sqrt{0^2 + \left(-\sqrt{7}\right)^2} = \sqrt{0 + 7} = \sqrt{\mathbf{7}}$.

3. $\left|\sqrt{15} - \sqrt{21}i\right| = \sqrt{15 + 21} = \sqrt{36} = \mathbf{6}$.

4. $|0| = |0 + 0i| = \sqrt{0^2 + 0^2} = \sqrt{0 + 0} = \sqrt{0} = \mathbf{0}$.

5. $\left|\sqrt{6} - \sqrt{11}i\right| = \sqrt{6 + 11} = \sqrt{\mathbf{17}}$.

Exercise 7.17:

1. $|(1 + i) - (1 - i)| = |2i| = \mathbf{2}$.

2. $|(2 - 3i) - (5 + 2i)| = |-3 - 5i| = \sqrt{(-3)^2 + (-5)^2} = \sqrt{9 + 25} = \sqrt{\mathbf{34}}$.

3. $|5 - (-i)| = |5 + i| = \sqrt{5^2 + 1^2} = \sqrt{25 + 1} = \sqrt{\mathbf{26}}$.

4. $|0 - (8 + 9i)| = |-8 - 9i| = \sqrt{(-8)^2 + (-9)^2} = \sqrt{64 + 81} = \sqrt{\mathbf{145}}$.

Exercise 7.19:

1. Center $= \mathbf{\textit{i}}$, Radius $= \mathbf{5}$

2. Center $= \mathbf{2 - 3\textit{i}}$, Radius $= \mathbf{11}$

3. Center $= \mathbf{-3 - 2\textit{i}}$, Radius $= \mathbf{2}$

4. Center $= \mathbf{0}$, Radius $= \mathbf{3}$

5. Center $= \mathbf{3 - \textit{i}}$, Radius $= \sqrt{\mathbf{17}}$

Exercise 7.22:

1. Center $= \mathbf{7 - 3\textit{i}}$, Radius $= \mathbf{4}$

2. Center $= \mathbf{1 - \textit{i}}$, Radius $= \mathbf{6}$

3. Center $= \mathbf{-2 + \textit{i}}$, Radius $= \sqrt{\mathbf{3}}$

Exercise 7.23:

1. $\{z \in \mathbb{C} \mid |z - 3i| \leq 1\}$ or $\{x + yi \in \mathbb{C} \mid x^2 + (y - 3)^2 \leq 1\}$

2. $\{z \in \mathbb{C} \mid |z - (8 + i)| \leq \sqrt{2}\}$ or $\{x + yi \in \mathbb{C} \mid (x - 8)^2 + (y - 1)^2 \leq 2\}$

3. $\{z \in \mathbb{C} \mid |z - (-1 + 2i)| \leq 1.5\}$ or $\{x + yi \in \mathbb{C} \mid (x + 1)^2 + (y - 2)^2 \leq 2.25\}$

Exercise 7.26:

1. **Open**

2. **Not open**

3. **Open**

4. **Not open**

5. **Open**

6. **Open**

7. **Open**

8. **Open**

9. **Open** (the definition of open is vacuously satisfied)

10. **Open**

Lesson 8

Exercise 8.3:

1. $\begin{bmatrix} 1 & -1 \\ -2 & 6 \end{bmatrix}$

2. $\begin{bmatrix} 3 & -1 \\ 5 & 5 \end{bmatrix}$

3. $\begin{bmatrix} -10 & 0 \\ -35 & 5 \end{bmatrix}$

4. $\begin{bmatrix} -4 & -2 \\ -25 & 15 \end{bmatrix}$

Exercise 8.4:

1. $m = 2, n = 2, F = \mathbb{Q}, \mathbb{R},$ or $\mathbb{C}$

2. $m = 2, n = 3, F = \mathbb{R}$ or $\mathbb{C}$

3. $m = 4, n = 1, F = \mathbb{C}$

4. $m = 1, n = 4, F = \mathbb{R}$ or $\mathbb{C}$

5. $m = 3, n = 4, F = \mathbb{Q}, \mathbb{R},$ or $\mathbb{C}$

Exercise 8.5:

1. $x_{23} = 3$

2. $x_{32} = 5$

3. $x_{34} = -6$

4. x_{43} **does not exist** (the matrix does **not** have 4 rows)

5. $x_{33} = 0$

6. $x_{24} = 4$

7. x_{42} **does not exist** (the matrix does **not** have 4 rows)

8. The size of the matrix is **3 × 4**.

Exercise 8.8:

1. $[1 \quad 3 \quad 4 \quad 0] \cdot \begin{bmatrix} 2 \\ 1 \\ 3 \\ 9 \end{bmatrix} = [1 \cdot 2 + 3 \cdot 1 + 4 \cdot 3 + 0 \cdot 9] = [2 + 3 + 12 + 0] = [17] = \mathbf{17}$.

2. $\begin{bmatrix} 2 \\ 1 \\ 3 \\ 9 \end{bmatrix} \cdot [1 \quad 3 \quad 4 \quad 0] = \begin{bmatrix} 2 & 6 & 8 & 0 \\ 1 & 3 & 4 & 0 \\ 3 & 9 & 12 & 0 \\ 9 & 27 & 36 & 0 \end{bmatrix}$.

3. $\begin{bmatrix} 1 & 2 \\ 3 & 4 \end{bmatrix} \cdot \begin{bmatrix} 5 & 6 \\ 0 & 1 \end{bmatrix} = \begin{bmatrix} 5+0 & 6+2 \\ 15+0 & 18+4 \end{bmatrix} = \begin{bmatrix} 5 & 8 \\ 15 & 22 \end{bmatrix}$.

4. $\begin{bmatrix} 5 & 6 \\ 0 & 1 \end{bmatrix} \cdot \begin{bmatrix} 1 & 2 \\ 3 & 4 \end{bmatrix} = \begin{bmatrix} 5+18 & 10+24 \\ 0+3 & 0+4 \end{bmatrix} = \begin{bmatrix} 23 & 34 \\ 3 & 4 \end{bmatrix}$.

Exercise 8.10:

1. Let $\begin{bmatrix} a & b \\ c & d \end{bmatrix}, \begin{bmatrix} e & f \\ g & h \end{bmatrix} \in M$. Then $a, b, c, d, e, f, g, h \in \mathbb{R}$. Since $\mathbb{R}$ is closed under addition, we have $a + e, b + f, c + g, d + h \in \mathbb{R}$. Therefore, $\begin{bmatrix} a & b \\ c & d \end{bmatrix} + \begin{bmatrix} e & f \\ g & h \end{bmatrix} = \begin{bmatrix} a+e & b+f \\ c+g & d+h \end{bmatrix} \in M$.

2. Let $\begin{bmatrix} a & b \\ c & d \end{bmatrix}, \begin{bmatrix} e & f \\ g & h \end{bmatrix}, \begin{bmatrix} i & j \\ k & l \end{bmatrix} \in M$. Since addition is associative in $\mathbb{R}$, we have

$$\left(\begin{bmatrix} a & b \\ c & d \end{bmatrix} + \begin{bmatrix} e & f \\ g & h \end{bmatrix} \right) + \begin{bmatrix} i & j \\ k & l \end{bmatrix} = \begin{bmatrix} a+e & b+f \\ c+g & d+h \end{bmatrix} + \begin{bmatrix} i & j \\ k & l \end{bmatrix} = \begin{bmatrix} (a+e)+i & (b+f)+j \\ (c+g)+k & (d+h)+l \end{bmatrix}$$

$$= \begin{bmatrix} a+(e+i) & b+(f+j) \\ c+(g+k) & d+(h+l) \end{bmatrix} = \begin{bmatrix} a & b \\ c & d \end{bmatrix} + \begin{bmatrix} e+i & f+j \\ g+k & h+l \end{bmatrix} = \begin{bmatrix} a & b \\ c & d \end{bmatrix} + \left(\begin{bmatrix} e & f \\ g & h \end{bmatrix} + \begin{bmatrix} i & j \\ k & l \end{bmatrix} \right).$$

3. Let $\begin{bmatrix} a & b \\ c & d \end{bmatrix}, \begin{bmatrix} e & f \\ g & h \end{bmatrix} \in M$. Since addition is commutative in $\mathbb{R}$, we have

$$\begin{bmatrix} a & b \\ c & d \end{bmatrix} + \begin{bmatrix} e & f \\ g & h \end{bmatrix} = \begin{bmatrix} a+e & b+f \\ c+g & d+h \end{bmatrix} = \begin{bmatrix} e+a & f+b \\ g+c & h+d \end{bmatrix} = \begin{bmatrix} e & f \\ g & h \end{bmatrix} + \begin{bmatrix} a & b \\ c & d \end{bmatrix}.$$

4. Let $\begin{bmatrix} a & b \\ c & d \end{bmatrix} \in M$ and let $\mathbf{0} = \begin{bmatrix} 0 & 0 \\ 0 & 0 \end{bmatrix}$. Since 0 is an identity in $\mathbb{R}$, we have

$$\begin{bmatrix} a & b \\ c & d \end{bmatrix} + \begin{bmatrix} 0 & 0 \\ 0 & 0 \end{bmatrix} = \begin{bmatrix} a+0 & b+0 \\ c+0 & d+0 \end{bmatrix} = \begin{bmatrix} a & b \\ c & d \end{bmatrix}$$

$$\begin{bmatrix} 0 & 0 \\ 0 & 0 \end{bmatrix} + \begin{bmatrix} a & b \\ c & d \end{bmatrix} = \begin{bmatrix} 0+a & 0+b \\ 0+c & 0+d \end{bmatrix} = \begin{bmatrix} a & b \\ c & d \end{bmatrix}.$$

It follows that $\mathbf{0}$ is an identity for M.

5. Let $\begin{bmatrix} a & b \\ c & d \end{bmatrix} \in M$. Since the additive inverse of $x \in \mathbb{R}$ is $-x$, we have

$$\begin{bmatrix} a & b \\ c & d \end{bmatrix} + \begin{bmatrix} -a & -b \\ -c & -d \end{bmatrix} = \begin{bmatrix} a + (-a) & b + (-b) \\ c + (-c) & d + (-d) \end{bmatrix} = \begin{bmatrix} 0 & 0 \\ 0 & 0 \end{bmatrix}$$

$$\begin{bmatrix} -a & -b \\ -c & -d \end{bmatrix} + \begin{bmatrix} a & b \\ c & d \end{bmatrix} = \begin{bmatrix} -a + a & -b + b \\ -c + c & -d + d \end{bmatrix} = \begin{bmatrix} 0 & 0 \\ 0 & 0 \end{bmatrix}.$$

It follows that $\begin{bmatrix} -a & -b \\ -c & -d \end{bmatrix}$ is an additive inverse of $\begin{bmatrix} a & b \\ c & d \end{bmatrix}$.

6. By parts 1 through 5 above, $(M, +)$ is a commutative group.

7. Let $k \in \mathbb{R}$ and let $\begin{bmatrix} a & b \\ c & d \end{bmatrix} \in M$. Then $a, b, c, d \in \mathbb{R}$. Since $\mathbb{R}$ is closed under multiplication, we have $ka, kb, kc, kd \in \mathbb{R}$. Therefore, $k \cdot \begin{bmatrix} a & b \\ c & d \end{bmatrix} = \begin{bmatrix} ka & kb \\ kc & kd \end{bmatrix} \in M$.

8. Let $\begin{bmatrix} a & b \\ c & d \end{bmatrix} \in M$. Then $1 \cdot \begin{bmatrix} a & b \\ c & d \end{bmatrix} = \begin{bmatrix} 1a & 1b \\ 1c & 1d \end{bmatrix} = \begin{bmatrix} a & b \\ c & d \end{bmatrix}$.

9. Let $j, k \in \mathbb{R}$ and let $\begin{bmatrix} a & b \\ c & d \end{bmatrix} \in M$. Since multiplication is associative in $\mathbb{R}$, we have

$$(jk) \begin{bmatrix} a & b \\ c & d \end{bmatrix} = \begin{bmatrix} (jk)a & (jk)b \\ (jk)c & (jk)d \end{bmatrix} = \begin{bmatrix} j(ka) & j(kb) \\ j(kc) & j(kd) \end{bmatrix} = j \begin{bmatrix} ka & kb \\ kc & kd \end{bmatrix} = j \left(k \begin{bmatrix} a & b \\ c & d \end{bmatrix} \right).$$

10. Let $k \in \mathbb{R}$ and let $\begin{bmatrix} a & b \\ c & d \end{bmatrix}, \begin{bmatrix} e & f \\ g & h \end{bmatrix} \in M$. Since multiplication is distributive over addition in $\mathbb{R}$, we have

$$k \left(\begin{bmatrix} a & b \\ c & d \end{bmatrix} + \begin{bmatrix} e & f \\ g & h \end{bmatrix} \right) = k \cdot \begin{bmatrix} a + e & b + f \\ c + g & d + h \end{bmatrix} = \begin{bmatrix} k(a + e) & k(b + f) \\ k(c + g) & k(d + h) \end{bmatrix}$$

$$= \begin{bmatrix} ka + ke & kb + kf \\ kc + kg & kd + kh \end{bmatrix} = \begin{bmatrix} ka & kb \\ kc & kd \end{bmatrix} + \begin{bmatrix} ke & kf \\ kg & kh \end{bmatrix} = k \cdot \begin{bmatrix} a & b \\ c & d \end{bmatrix} + k \cdot \begin{bmatrix} e & f \\ g & h \end{bmatrix}$$

11. Let $j, k \in \mathbb{R}$ and let $\begin{bmatrix} a & b \\ c & d \end{bmatrix} \in M$. Since multiplication is distributive over addition in $\mathbb{R}$, we have

$$(j + k) \begin{bmatrix} a & b \\ c & d \end{bmatrix} = \begin{bmatrix} (j + k)a & (j + k)b \\ (j + k)c & (j + k)d \end{bmatrix} = \begin{bmatrix} ja + ka & jb + kb \\ jc + kc & jd + kd \end{bmatrix}$$

$$= \begin{bmatrix} ja & jb \\ jc & jd \end{bmatrix} + \begin{bmatrix} ka & kb \\ kc & kd \end{bmatrix} = j \cdot \begin{bmatrix} a & b \\ c & d \end{bmatrix} + k \cdot \begin{bmatrix} a & b \\ c & d \end{bmatrix}.$$

12. By parts 6 through 11 above, M is a vector space over $\mathbb{R}$.

INDEX

About the Author

Dr. Steve Warner, a New York native, earned his Ph.D. at Rutgers University in Pure Mathematics in May 2001. While a graduate student, Dr. Warner won the TA Teaching Excellence Award.

After Rutgers, Dr. Warner joined the Penn State Mathematics Department as an Assistant Professor and in September 2002, he returned to New York to accept an Assistant Professor position at Hofstra University. By September 2007, Dr. Warner had received tenure and was promoted to Associate Professor. He has taught undergraduate and graduate courses in Precalculus, Calculus, Linear Algebra, Differential Equations, Mathematical Logic, Set Theory, and Abstract Algebra.

From 2003 – 2008, Dr. Warner participated in a five-year NSF grant, "The MSTP Project," to study and improve mathematics and science curriculum in poorly performing junior high schools. He also published several articles in scholarly journals, specifically on Mathematical Logic.

Dr. Warner has nearly two decades of experience in general math tutoring and tutoring for standardized tests such as the SAT, ACT, GRE, GMAT, and AP Calculus exams. He has tutored students both individually and in group settings.

In February 2010 Dr. Warner released his first SAT prep book "The 32 Most Effective SAT Math Strategies," and in 2012 founded Get 800 Test Prep. Since then Dr. Warner has written books for the SAT, ACT, SAT Math Subject Tests, AP Calculus exams, and GRE. In 2018 Dr. Warner released his first pure math book called "Pure Mathematics for Beginners." Since then he has released several more books, each one addressing a specific subject in pure mathematics.

Dr. Steve Warner can be reached at

steve@SATPrepGet800.com

BOOKS BY DR. STEVE WARNER

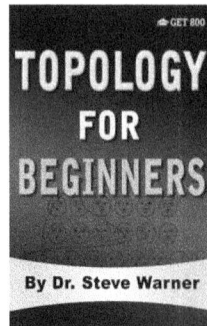

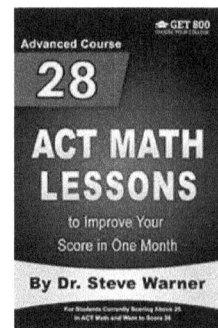

PURE MATHEMATICS FOR BEGINNERS
Logic
Set Theory
Abstract Algebra
Number Theory
Real Analysis
Topology
Complex Analysis
Linear Algebra
By Dr. Steve Warner

SET THEORY FOR BEGINNERS
By Dr. Steve Warner

TOPOLOGY FOR BEGINNERS
By Dr. Steve Warner

ABSTRACT ALGEBRA FOR BEGINNERS
By Dr. Steve Warner

48 NEW SAT MATH LESSONS to Improve Your Score in 48 Days
By Dr. Steve Warner

Beginner Course 28 NEW SAT MATH LESSONS to Improve Your Score in One Month
By Dr. Steve Warner
For Students Currently Scoring Below 500 in SAT Math

Intermediate Course 28 NEW SAT MATH LESSONS to Improve Your Score in One Month
By Dr. Steve Warner
For Students Currently Scoring Between 500 and 600 in SAT Math

Advanced Course 28 NEW SAT MATH LESSONS to Improve Your Score in One Month
By Dr. Steve Warner
For Students Currently Scoring Above 600 in SAT Math and Want to Score 800

500 NEW SAT MATH PROBLEMS arranged by Topic and Difficulty Level
By Dr. Steve Warner

Second Edition 320 SAT MATH PROBLEMS arranged by Topic and Difficulty Level
By Dr. Steve Warner

LEVEL 1 320 SAT MATH SUBJECT TEST PROBLEMS arranged by Topic and Difficulty Level
By Dr. Steve Warner

LEVEL 2 320 SAT MATH SUBJECT TEST PROBLEMS arranged by Topic and Difficulty Level
By Dr. Steve Warner

Second Edition 320 ACT MATH PROBLEMS arranged by Topic and Difficulty Level
By Dr. Steve Warner

Beginner Course 28 ACT MATH LESSONS to Improve Your Score in One Month
By Dr. Steve Warner

Intermediate Course 28 ACT MATH LESSONS to Improve Your Score in One Month
By Dr. Steve Warner

Advanced Course 28 ACT MATH LESSONS to Improve Your Score in One Month
By Dr. Steve Warner

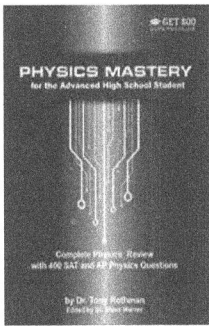

PHYSICS MASTERY
for the Advanced High School Student

Complete Physics Review
with 400 SAT and AP Physics Questions

by Dr. Tony Rothman
Edited by Dr. Steve Warner

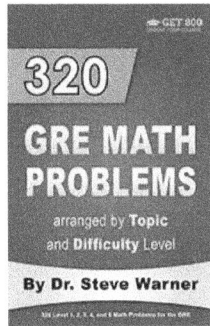

GET 800

320
GRE MATH PROBLEMS

arranged by Topic
and Difficulty Level

By Dr. Steve Warner

320 Level 1, 2, 3, 4, and 5 Math Problems for the GRE

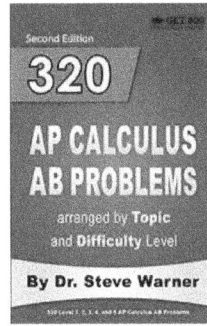

Second Edition

GET 800

320
AP CALCULUS AB PROBLEMS

arranged by Topic
and Difficulty Level

By Dr. Steve Warner

320 Level 1, 2, 3, 4, and 5 AP Calculus AB Problems

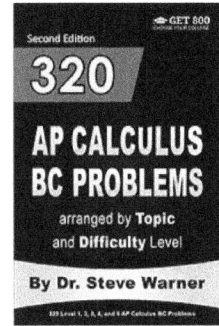

Second Edition

GET 800

320
AP CALCULUS BC PROBLEMS

arranged by Topic
and Difficulty Level

By Dr. Steve Warner

320 Level 1, 2, 3, 4, and 5 AP Calculus BC Problems

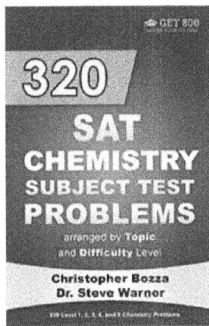

GET 800

320
SAT CHEMISTRY SUBJECT TEST PROBLEMS

arranged by Topic
and Difficulty Level

Christopher Bozza
Dr. Steve Warner

320 Level 1, 2, 3, 4, and 5 Chemistry Problems

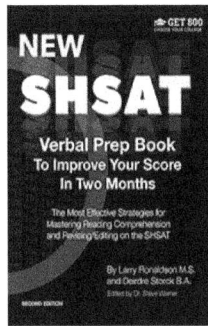

NEW
SHSAT
Verbal Prep Book
To Improve Your Score
In Two Months

The Most Effective Strategies for
Mastering Reading Comprehension
and Revising/Editing on the SHSAT

By Larry Ronaldson M.S.
and Deirdre Storck B.A.

Edited by Dr. Steve Warner

SECOND EDITION

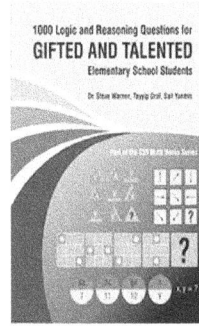

1000 Logic and Reasoning Questions for
GIFTED AND TALENTED
Elementary School Students

Dr. Steve Warner, Tayyip Oral, Sali Yannis

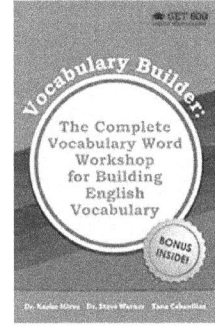

GET 800

Vocabulary Builder:
The Complete
Vocabulary Word
Workshop
for Building
English
Vocabulary

BONUS
INSIDE!

Dr. Kaulas SGree   Dr. Steve Warner   Tana Cebunllas

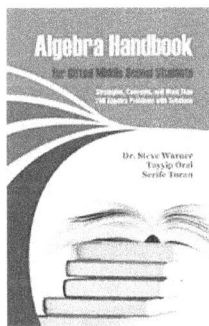

Algebra Handbook
for Gifted Middle School Students

Strategies, Concepts, and More Than
200 Algebra Problems with Solutions

Dr. Steve Warner
Tayyip Oral
Serife Turan

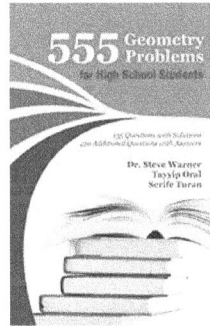

555 Geometry Problems
for High School Students

135 Questions with Solutions
250 Additional Questions with Answers

Dr. Steve Warner
Tayyip Oral
Serife Turan

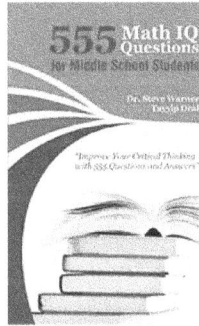

555 Math IQ Questions
for Middle School Students

Dr. Steve Warner
Tayyip Oral

"Improve Your Critical Thinking
with 555 Questions and Answers"

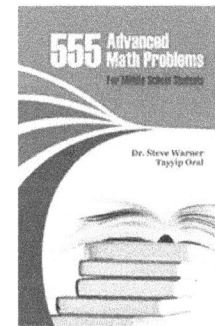

555 Advanced Math Problems
for Middle School Students

Dr. Steve Warner
Tayyip Oral

www.ingramcontent.com/pod-product-compliance
Lightning Source LLC
Chambersburg PA
CBHW081814200326

41597CB00023B/4251